196

Advances in Polymer Science

Advances in Polymer Science

Recently Published and Forthcoming Volumes

Supramolecular Polymers/Oligomers/ Polymeric Betains
Vol. 201, 2006

Ordered Polymeric Nanostructures at Surfaces
Materials, Chemistry, Formation, Properties, and Applications
Volume Editor: Vansco, G. J.
Vol. 200, 2006

Emissive Materials/Nanomaterials
Vol. 199, 2006

Surface-Initiated Polymerization II
Volume Editor: Jordan, R.
Vol. 198, 2006

Surface-Initiated Polymerization I
Volume Editor: Jordan, R.
Vol. 197, 2006

Conformation-Dependent Design of Sequences in Copolymers II
Volume Editor: Khokhlov, A. R.
Vol. 196, 2006

Conformation-Dependent Design of Sequences in Copolymers I
Volume Editor: Khokhlov, A. R.
Vol. 195, 2006

Enzyme-Catalyzed Synthesis of Polymers
Volume Editors: Kobayashi, S., Ritter, H., Kaplan, D.
Vol. 194, 2006

Polymer Therapeutics II
Polymers as Drugs, Conjugates and Gene Delivery Systems
Volume Editors: Satchi-Fainaro, R., Duncan, R.
Vol. 193, 2006

Polymer Therapeutics I
Polymers as Drugs, Conjugates and Gene Delivery Systems
Volume Editors: Satchi-Fainaro, R., Duncan, R.
Vol. 192, 2006

Interphases and Mesophases in Polymer Crystallization III
Volume Editor: Allegra, G.
Vol. 191, 2005

Block Copolymers II
Volume Editor: Abetz, V.
Vol. 190, 2005

Block Copolymers I
Volume Editor: Abetz, V.
Vol. 189, 2005

Intrinsic Molecular Mobility and Toughness of Polymers II
Volume Editor: Kausch, H.-H.
Vol. 188, 2005

Intrinsic Molecular Mobility and Toughness of Polymers I
Volume Editor: Kausch, H.-H.
Vol. 187, 2005

Polysaccharides I
Structure, Characterization and Use
Volume Editor: Heinze, T.
Vol. 186, 2005

Advanced Computer Simulation Approaches for Soft Matter Sciences II
Volume Editors: Holm, C., Kremer, K.
Vol. 185, 2005

Crosslinking in Materials Science
Vol. 184, 2005

Conformation-Dependent Design of Sequences in Copolymers II

Volume Editor: Alexei R. Khokhlov

With contributions by

V. O. Aseyev · A. Y. Grosberg · A. R. Khokhlov
S. I. Kuchanov · V. I. Lozinsky · H. Tenhu · F. M. Winnik

Springer

The series *Advances in Polymer Science* presents critical reviews of the present and future trends in polymer and biopolymer science including chemistry, physical chemistry, physics and material science. It is adressed to all scientists at universities and in industry who wish to keep abreast of advances in the topics covered.
As a rule, contributions are specially commissioned. The editors and publishers will, however, always be pleased to receive suggestions and supplementary information. Papers are accepted for *Advances in Polymer Science* in English.
In references *Advances in Polymer Science* is abbreviated *Adv Polym Sci* and is cited as a journal.

Springer WWW home page: http://www.springer.com
Visit the APS content at http://www.springerlink.com/

ISSN 0065-3195
ISBN-10 3-642-43948-9 Springer Berlin Heidelberg New York
ISBN-13 978-3-642-43948-3 Springer Berlin Heidelberg New York
DOI 10.1007/11570325

Springer is a part of Springer Science+Business Media

springer.com

Softcover re-print of the Hardcover 1st edition 2006

Cover design: *Design & Production* GmbH, Heidelberg
Typesetting and Production: LE-TEX Jelonek, Schmidt & Vöckler GbR, Leipzig

Printed on acid-free paper 02/3141 YL – 5 4 3 2 1 0

Volume Editor

Advances in Polymer Science Also Available Electronically

For all customers who have a standing order to Advances in Polymer Science, we offer the electronic version via SpringerLink free of charge. Please contact your librarian who can receive a password or free access to the full articles by registering at:

springerlink.com

If you do not have a subscription, you can still view the tables of contents of the volumes and the abstract of each article by going to the SpringerLink Homepage, clicking on "Browse by Online Libraries", then "Chemical Sciences", and finally choose Advances in Polymer Science.

You will find information about the

- Editorial Board
- Aims and Scope
- Instructions for Authors
- Sample Contribution

at springeronline.com using the search function.

Advances in Polymer Science
Also Available Electronically

Preface

For a long time, chemical industry was focused on polymers mainly from the viewpoint of obtaining advanced construction materials, such as plastics, rubbers, fibers, polymer composites. These materials provided a number of important benefits, including improved strength and long-term durability, light weight, environmental resistance, and design flexibility. Starting from about the early 1980s, the main focus of interest shifted to functional polymers. Among these are superabsorbents, nanoporous rate-controlling membranes, reversible adhesives, electro-conductive polymers and nanowires. In the 1990s, the scientific and industrial polymer community started to discuss "smart" or "intellectual" polymer systems (e.g., soft manipulators, polymer systems for controlled drug release, field-responsive polymers, shape memory networks, and self-healing coatings); the meaning behind these terms is that simply the functions performed by polymers become more sophisticated and diverse. The line of research concentrating on polymer systems with more and more complex functions will certainly be in the mainstream of polymer science in the 21st century.

One of the ways to obtain new polymers for sophisticated functions is connected to the synthesis of novel building blocks – monomer species – where the required function is linked to the chemical structure of these blocks. However, the potential of this approach is rather limited because complicated and diverse functions of polymeric materials would then require a very complex structure of monomers, which normally means that the organic synthesis is more expensive and less robust. An alternative approach is to use known building blocks and to try to design a copolymer with a given sequence of these units. At the present time, there are many synthetic and theoretical strategies directed towards varying the chemical sequences of copolymers: from the variation of their composition and blocky structure to more sophisticated features like "tapered" and gradient structures. In a broad sense, most conventional chemical syntheses, especially polymer syntheses, can be regarded as *bottom-up nanotechnology* leading to the assembly of building blocks into the final macromolecules. Unfortunately, in most conventional polymerization processes, the physical control of assembly during the reactions is practically impossible.

With these difficulties in mind, it is instructive to look at main biological macromolecules – proteins, DNA, and RNA – that have precise and specific structures. These polymers in living systems are responsible for functions, which are incomparably more complex and diverse than the functions that we normally discuss for synthetic copolymers. The molecular basis for this ability to perform sophisticated functions is associated with the primary sequences of biopolymers. In particular, the unique functions of proteins reflect their molecular structure, so that the sequence of amino acids in a protein defines its secondary and tertiary structure (fibrous versus globular, for example) as well as its function. Therefore, the study of biopolymers at the molecular level may point to new directions in materials design and construction – not just actually using biomolecules themselves to construct novel materials, but mimicking the specific primary sequences of biological polymers. Indeed, by taking this biomimetic route, significant and path-breaking achievements may be made by materials scientists and engineers. Thus, a promising path related to the design of new synthetic copolymers is to learn the rules of biologically driven controlled synthesis and directed assembly in vivo and eventually to apply these rules to the creation of engineered synthetic polymer systems from cheap and commonly available building blocks.

Original ideas connected with the biomimetic design of sequences in synthetic functional copolymers were formulated by us in 1998. They were based on the simple and well-known fact that the function of all globular proteins depends on two main factors: (1) they are globular and (2) they are soluble in aqueous medium. The combination of these two factors is nontrivial, e.g., for homopolymers and random copolymers the transition to globular conformation is usually accompanied by the precipitation of globules from the solution. Protein globules are soluble in water because of the special primary sequence: in the native conformation most of hydrophobic monomer units are in the core of the globule while hydrophilic monomer units form the envelope of this core. Keeping the biomimetic approach described above in mind, one can formulate the following problem: is it possible to design such a sequence of synthetic HP copolymer (copolymer consisting of monomer units of two types, H and P) that, in the most dense globular conformation, all hydrophobic H-units are in the core of this globule while the hydrophilic (polar) P-units form the envelope of this core? The corresponding bio-inspired two-letter HP copolymers generated on a computer were called *protein-like* copolymers. Of course, the degree of function complexity that we hope to achieve for designed copolymers, is much less than current biopolymers, but the behavior of copolymers with designed sequences can exhibit many useful features distinguishing them from the "scratch" (e.g., statistically random) sequences.

The approach formulated in 1998 is based on the assumption that a copolymer obtained under some preparation conditions is able to "remember" features of its original conformation from which it was built and to store the corresponding information in the resulting sequence. Therefore, this approach may be called *conformation-dependent sequence design*.

Most of the contributions collected in volumes 195 and 196 are dedicated to the review of the results obtained recently in this direction, i.e., dealing with the conformation-dependent sequence design of copolymers and the study of their properties. The contributions collected present the up-to-date research results on most of the topics related to this approach. Surface and interfacial phenomena are also major topics of these volumes. The individual chapters are diverse in purpose and in style. Some of them paint the field in broad, conceptual strokes, others in fine methodical detail. Some present information, others arguments or interpretation. Some summarize past activities, others point to future potential.

Volume 195 begins with a chapter by Khalatur and Khokhlov that gives a survey of the simulation methods as applied to the design of nontrivial sequences in synthetic copolymers aimed to achieve desired functional properties. Several new synthetic strategies allowing for the synthesis of copolymers with a broad variation of their sequence distributions are reported. Synthetic copolymers exhibiting long-range statistical correlations, large-scale compositional heterogeneities, and physical complexity are the focal point of this review. Roughly speaking, the physical complexity of a sequence is understood as the amount of information that is stored in that sequence about a particular environment.

Zhang and Wu review the experimental results on the folding of different hydrophilically or hydrophobically modified copolymers in extremely dilute solutions. The focus is on the formation of stable and soluble mesoglobules made of several amphiphilic copolymers and on the folding of single copolymers – both linear and grafted – into core-shell nanostructures that have various applications. All these features are directly related to the unique chemical sequence of the copolymer chains under discussion. The authors discuss the insights that can be obtained by the analysis of the data of laser light scattering and the use of this method as a potential probe of fine molecular structures, including the so-called "molten" globule state. Emphasis is put on the most recent achievements, although important historical contributions are also mentioned.

A concept of amphiphilicity, as applied to single monomer units of designed water-soluble polymers, is presented in the third chapter by Okhapkin, Makhaeva, and Khokhlov. The concept is relevant to biomolecular structures and assemblies in aqueous solution. The authors consider the substantial body of information obtained experimentally and theoretically on surface molecular chemical structures, including those that are prospective for surface catalysis. Unusual conformational behaviors of single amphiphilic polymers recently observed in simulations are also discussed in detail.

The problems related to the colloidal stability of amphiphilic polymers in water are reviewed by Aseyev, Tenhu, and Winnik in the first chapter of volume 196. The focus is on the derivatives of thermally responsive smart macromolecules – both on copolymers and homopolymers – which are present in a solution as stable micelles potentially having various applications.

One of the promising synthetic strategies of conformation-dependent sequence design is based on direct copolymerization under unusual conditions.

This strategy was first realized by Lozinsky et al., who studied the redox-initiated free-radical copolymerization of thermosensitive N-vinylcaprolactam with hydrophilic N-vinylimidazole at different temperatures, as well as by Chi Wu and coworkers. Lozinsky presents an extensive review of the experimental approaches, both already described in the literature and potential new ones, to chemical synthesis of protein-like copolymers capable of forming core-shell nanostructures in a solution.

Continuing along the chemical theme, Kuchanov and Khokhlov review comprehensively the diffusion-controlled polymer-analogous reactions and free-radical copolymerization. The field of polymer chemistry today is distinguished by its depth of mathematical and quantitative rigor in the pursuit of a wide array of challenging new subjects. These ingredients are present in full in this chapter, where the quantitative theory of solution and interphase free-radical copolymerization is discussed from the viewpoint of statistical chemistry and statistical physics of polymers. It is shown that the interaction of these two disciplines has significant impact on the development of new synthetic strategies and technologies in polymer chemistry, including those related to the conformation-dependent design of nontrivial copolymer sequences.

The final short historical review by Grosberg and Khokhlov is devoted exclusively to the discussion of fundamental ideas initially formulated in polymer physics by the outstanding Russian scientist I.M. Lifshitz and especially to today's development of his ideas. Major attention is focused on the statistical theory of heteropolymers covering such areas as protein folding and the sequence design both of protein macromolecules and of synthetic copolymers mimicking some protein properties. In particular, the ideas of sequence design in functional copolymers were originated within the school of Lifshitz; thus, in volume 196, it seems natural to give a review of the achievements along the lines of all major ideas initially formulated by Lifshitz in the field of polymer science.

All of the selected contributions that are present in these special volumes are good representatives for manifesting the importance of the concepts based on conformation-dependent sequence design. It has been our intention to provide the scientific and industrial polymer community with a comprehensive view of the current state of knowledge on designed polymers. Both volumes attempt to review what is currently known about these polymers in terms of their synthesis, chemical and physical properties, and applications. We will feel the volumes have been successful if some of the chapters presented here stimulate readers to become interested in and solve specific problems in this rapidly developing field of research.

Moscow, November 2005 *Alexei R. Khokhlov*

Contents

Contents of Volume 195

Conformation-Dependent Design of Sequences in Copolymers I

Volume Editor: Alexei R. Khokhlov
ISBN: 3-540-29513-5

Adv Polym Sci (2006) 196: 1–85
DOI 10.1007/12_052

Published online: 6 December 2005

Temperature Dependence of the Colloidal Stability of Neutral Amphiphilic Polymers in Water

Vladimir O. Aseyev[1] (✉) · Heikki Tenhu[1] · Françoise M. Winnik[2]

[1]Laboratory of Polymer Chemistry, PB 55, University of Helsinki, 00014 HY Helsinki, Finland
vladimir.aseyev@helsinki.fi, heikki.tenhu@helsinki.fi

[2]Department of Chemistry and Faculty of Pharmacy, University of Montreal, CP 6128 succursale Centre-Ville, Montreal, QC H3C 3J7, Canada
francoise.winnik@umontreal.ca

Abstract Various thermally responsive polymers and their derivatives form colloidally stable nanosized particles upon heating of their aqueous solutions. A compilation of the types of polymers reported to exhibit this behaviour is presented in this review, including poly(*N*-isopropylacrylamide), poly(*N*-vinylcaprolactam), and poly(vinyl methyl ether) together with their copolymers with either hydrophobic or hydrophilic monomers. In all cases experimental evidence is presented, together with various theoretical models. Possible mechanisms responsible for the stabilisation of these colloids are discussed.

Keywords Colloidal stability · Globule · Mesoglobule · Thermoresponsive polymers

Abbreviations

A_2	Osmotic second virial coefficient
A_3	Osmotic third virial coefficient
AIBN	2,2′-Azobis(isobutyronitrile)
c	Polymer concentration
C_p	Partial heat capacity
DLS	Dynamic light scattering
DLVO	Derjaguin–Landau–Verwey–Overbeek
DSC	Differential scanning calorimetry
EO	Ethylene oxide
GMA	Glycidylmethacrylate
HS-DSC	High-sensitivity differential scanning calorimetry
KPS	Potassium persulphate
LCST	Lower critical solution temperature
$MAC_{11}EO_{42}$	ω-Methoxy poly(ethylene oxide)$_{42}$ undecyl α-methacrylate
M_w	Weight-average molar mass
NASI	*N*-Acryloylsuccinimide
NIPAM	*N*-Isopropylacrylamide
Np	Naphthalene
NRET	Nonradiative energy transfer
PEO	Poly(ethylene oxide)
PNIPAM	Poly(*N*-isopropylacrylamide)
PNIPMAM	Poly(*N*-isopropylmethacrylamide)
PPC	Pressure perturbation calorimetry
PS	Polystyrene
P*t*BMA	Poly(*tert*-butyl methacrylate)
PVCL	Poly(*N*-vinylcaprolactam)
PVME	Poly(vinyl methyl ether)
PVP	Poly(*N*-vinylpyrrolidone)
Py	Pyrene
R_g	Mean-square radius of gyration
R_h	Hydrodynamic radius
SDS	Sodium dodecyl sulphate
St	Styrene
T	Temperature
t	Time
T_{cp}	Cloud-point temperature
T_{dem}	Demixing temperature
T_g	Glass-transition temperature
T_{max}	Temperature of maximum heat capacity
TM-DSC	Temperature-modulated differential scanning calorimetry
U	Internal energy of segmental interactions
UCST	Upper critical solution temperature
VCL	*N*-Vinylcaprolactam

VP	Vinylpyrrolidone
ΔH	Heat of transition
ΔT	Shock cooling depth
$\Delta T_{1/2}$	Width of the thermogram of the transition at half-height
α_{pol}	Thermal expansion coefficient of a polymer
η_{sp}/c	Reduced viscosity
ρ	Average density
ζ	Zeta potential

1 Introduction

1.1 Definitions of the Colloidal Stability

Throughout this review, the terms stability, metastability, and stable particles will be employed profusely. They need to be defined at the onset, in order to prevent possible misunderstandings. A given system may exist in two stable states: the thermodynamically stable state and the colloidally stable state. According to the IUPAC Recommendations [1], a system under a given set of conditions is thermodynamically metastable if it is in a state corresponding to a local minimum of the appropriate thermodynamic potential, such as the Gibbs energy, for specified constraints imposed upon the system, constant temperature and pressure for instance. If this system can exist in several states, the state of lowest free energy is called the thermodynamically stable state. Any system, if left alone, finally adopts the stable state. The time required for that process to occur is determined by the magnitude of the activation energy barriers which separate the metastable and stable states.

For polymer solutions, a decrease of the solvent thermodynamic quality tends to decrease polymer–solvent interactions and increase polymer–polymer interactions. This results in intermolecular association and subsequent macrophase separation. Several strategies can be implemented to avoid intermolecular contacts. One may employ dilute polymer solutions to decrease the rate of precipitation. Another possibility is to modify the surface of the interpolymeric aggregates with nonadhesive or repulsive groups. Such intentionally stabilised particles should not be confused with the so-called colloidally stable particles. The latter term refers to particles that do not aggregate at a significant rate in a thermodynamically unfavourable medium [1]. It is usually employed to describe systems which do not phase-separate on the macroscopic level during the time of an experiment. Typical polymeric colloidally stable particles range in size from approximately 1 nm to approximately 1 μm. They adopt various shapes: fibres, thin films, spheres, porous solids, gels, etc. According to our definitions, colloidally stable particles are thermodynamically metastable and they may remain in this metastable state for a long time, depending on the affinity of the polymer with its medium.

The challenge of achieving colloidal stability of aqueous systems was addressed already in antiquity. More than 4000 years ago, in ancient Egypt, a method was developed to stabilise an ink composition by adding hydrophilic polysaccharides obtained from acacias to a colloidal (unstable) dispersion of soot particles [2]. The hydrophilic polymer adsorbed on the particles and the ink remained stable over a long period of time. Nowadays, colloid chemists employ several strategies to enhance the stability of aqueous polymer dispersions. In general, the methods can be divided into two broad categories, depending on whether they involve electrostatic or steric stabilisation. Electrostatic stabilisation, which is based on the repulsion of the electric double layer of the particles, is typically realised via the use of ionic initiators and ionic surfactants, or via copolymerisation with monomers having groups capable of dissociating in water. Such particles repel each other owing to the entropic (osmotic) pressure caused by the counterions between the surfaces. The interactions among charged surfaces are usually explained by the Derjaguin–Landau–Verwey–Overbeek (DLVO) theory, which combines short-range attractive van der Waals forces and long-range electrostatic double-layer forces. The strength of the repulsion force between the particles and the thickness of the electric double layer may be altered by changing the ionic strength of the aqueous medium [3]. At high electrolyte concentrations, the repulsion between the particles vanishes and the coagulation of the particles is fully diffusion controlled [4, 5].

Steric stabilisation may be achieved by decorating the particle surface with neutral water soluble polymers, such as poly(ethylene oxide) (PEO). The repulsive force has an entropic origin: when two grafted surfaces approach each other they experience a repulsive force once the grafted chains begin to overlap and the mobility of the chains decreases [3, 4]. Steric stabilisation by nonionic hydrophilic polymers is typically independent of the ionic strength, assuming that the added electrolyte does not drastically change the thermodynamic quality of the aqueous solvent. PEO has been shown to be an effective steric stabiliser, even at high electrolyte concentrations as long as the molecular mass of PEO is sufficiently high [6]. The use of PEO is particularly advantageous in the case of colloids designed for biotechnological applications, since PEO considerably prevents the adsorption of proteins onto polymer surfaces and, thus, increases the biocompatibility of the polymer [7].

Stabilising repulsive forces may also arise from specific chemical and/or physical properties of the particle surface. Repulsion might be caused by the thermal fluctuation forces between two undulating surfaces, as in the case of lipid bilayers where a repulsive short-range force has been observed [8]. Steric repulsion originates from hydrated molecules that protrude somewhat into the aqueous phase owing to thermal motion [8]. It is often suggested that the high hydrophilicity of a surface may lead to interparticle repulsion, even if the surface does not possess any electric charge or repulsive polymer layer. This type of repulsion has been ascribed to a force, the hydration or structural force, which arises as a consequence of the specific structure of the hydrogen-bonded water layer on the particle surface. It has been suggested that the overlap of two structurally modified boundary layers gives rise to hydrophilic repulsion [9–11]. The existence of such hydration forces remains a subject of controversy. It has been argued that in most cases the non-DLVO forces observed in the short-range region may simply be repulsive forces, such as the undulation or protruding forces mentioned previously, especially when the surface is rough [12, 13].

1.2 Definitions of Neutral Amphiphilic Polymers

The physical properties of water determine the solubility, conformation, and the subsequent reactions of synthetic amphiliphilic polymers, as they do in the case of the biopolymers responsible for life on earth. In spite of numerous studies, the detailed structure of water remains the subject of controversy [14–26]. The key properties of water as a solvent are its high polarity and high dielectric constant, and the high level of organisation due to the formation of hydrogen bonds between water molecules. Compounds are expected to dissolve in water if they possess polar groups or/and groups capable of dissociating or of forming hydrogen bonds. Solutes introduced into water disrupt the local structure and create new structures. Thus, ions such as Na^+, Cl^-, SO_4^{2-}, or PO_4^{3-} orient water molecules by ion-dipole forces, whereas hydrophilic nonionic solutes, such as acetone, form hydrogen bonds with water molecules, more or less compensating the broken water–water hydrogen bonds. One tends to associate the term "hydrophobic" with a substance or a functional group of a polymer that "dislikes" water. In fact, the interaction between a hydrophobic moiety and water is attractive, owing to dispersion forces. Hydrophobic solutes, which are unable to form hydrogen bonds with water, cause the formation of local regions of "hydrophobic hydration", where the water molecules, which surround the solute, are even more ordered and more hydrogen-bonded than those in the bulk. This feature of hydrophobic molecules in water gives rise to strong solvent-mediated attraction between hydrophobic molecules in water, the so-called hydrophobic effect. It triggers the release of some of the highly structured water molecules into bulk water. The release of these water molecules changes the enthalpy of the solution and, more importantly, considerably increases the entropy of the system. At room temperature, the entropy change prevails over the enthalpy change. Therefore the "hydrophobic effect" process is considered to be "entropically driven" [27–40].

Various models have been proposed to describe the interaction between hydrophobic solutes and water [19, 20, 41–45]. In a recent publication, Moelbert and De Los Rios [45] proposed a unified approach to describe analytically the upper and lower critical solution behaviour of aqueous polymers in water. They discussed earlier models and developed a mean-field theory adapted from the Muller–Lee–Graziano model [19, 20]. The model separates water molecules into two different populations, on the basis of the number of hydrogen bonds they form. Water molecules that are highly hydrogen-bonded to their neighbours have fewer rotational degrees of freedom and thus a lower multiplicity of degenerate configurations (lower entropy), compared with unbound molecules. The appropriate description of the solvent in the vicinity of hydrophobic solute particles, which may be much larger than individual water molecules, is obtained by allowing each site of a lattice representing the system to be occupied by a group of water molecules. The bimodal nature of the Muller–Lee–Graziano model is preserved in this Moelbert–De Los Rios version by specifying only two types of water clusters at each site, where an "ordered" site is one in which most of the hydrogen bonds between the molecules in the cluster are intact, while a "disordered" site is one for which a number of hydrogen bonds are broken. The Muller–Lee–Graziano approach describes the energy levels of water molecules as a function of their proximity to nonpolar solute molecules. It provides an appropriate description of the many-body interactions between the hydrophobic solute particles. The solubility and aggregation of hydrophobic substances were studied by evaluating detailed Monte Carlo simulations in the vicinity of the first-order aggregation phase transition.

The hydrophilicity of a polymer and, therefore, its ability to mix with water arises from specific interactions between its functional groups and water molecules. Polyelectrolytes are soluble in water owing to the presence along the chain of highly hydrated ionic units. Their behaviour in water is rather well understood. Many neutral polymers, such as acrylic polymers that carry amide, ether, or alcohol groups, readily dissolve in water. Their solubility can be traced to the formation of hydrogen bonds between hydrophilic groups and water molecules, which contributes favourably to the free energy of mixing. Hydrophilic groups, which are capable of forming hydrogen bonds with water, typically, but not always, are attached to side chains, whereas the hydrocarbon backbone as such is insoluble in water. The entire polymer is solubilised in aqueous media if the number of hydrophilic groups along the polymer chain is high enough. Water molecules, which are incapable of forming hydrogen bonds with nonpolar groups, reorient themselves around these nonpolar groups, forming regions of structured water and thus decreasing the entropy upon mixing (the hydrophobic effect) [31, 33]. Hydrophobic groups have a tendency to associate in order to decrease the overall surface area of interaction with water and release structured water molecules in bulk. At higher temperatures, the entropy term dominates over the exothermic enthalpy of hydrogen-bond formation. Therefore, the free-energy change upon mixing becomes positive and phase separation takes place. The replacement of polymer–water contacts with polymer–polymer and water–water contacts manifests itself macroscopically by precipitation. The detailed aspects of the solubility of neutral water-soluble polymers are not well understood. For a comprehensive description of various water-soluble polymers and their properties, the reader is referred to Refs. [31, 34, 36, 37, 40] and the *CRC handbook of thermodynamic data of aqueous polymer solutions* [46].

Water-soluble polymers always contain hydrophobic moieties, in addition to ionic or hydrogen-bond active groups. They are amphiphilic and in water they experience both repulsive and attractive forces. It is the delicate balance between these forces that determines the solubility of the macromolecule in water. This balance is easily affected by various environmental stimuli, for example electromagnetic field, ionic strength, pH, or temperature. Changes of such external factors affect the solubility of a macromolecule, triggering conformational transformations and eventually cause phase separation. Note that a conformational coil–globule transition and phase-separation can also be achieved by adding a nonsolvent to an aqueous polymer solution. A polymer chain dissolved in a good solvent shrinks when a certain amount of nonsolvent is added to the solution [47–49]. Over the last few decades the study of stimuli-responsive polymers, sometimes called intelligent or smart polymers, has blossomed, not only from the theoretical viewpoint, but also for their practical applications in many areas of materials science. Recommended further reading on stimuli-responsive water-soluble and amphiphilic polymers includes a monograph [50] and a recent review on stimuli-responsive polymers and their bioconjugates [51].

1.3 Effect of Temperature on Polymer Solutions

Typically, a polymer responds to changes in its solution temperature by altering its conformation, for example by undergoing a coil–globule transition. The first fundamental analysis of this phenomenon is that of Flory, who laid down the basis of our current understanding. According to Flory [52, 53], the swelling, in a given solvent, of a single neutral linear macromolecule consisting of structurally simple repeat units is determined by two main factors: the excluded-volume interactions and the elastic forces. If the thermal energy $k_B T$ of the repeat units is high, the excluded-volume interactions prevail over the attraction between repeating units and, consequently, the macromolecule swells. This is the case for a thermodynamically good solvent, in which a linear homopolymer adopts the conformation of a very loose extended coil. The constraints limiting its expansion are the C – C covalent bonds and the entropy of the coil, which decreases with coil expansion owing to the decrease in the number of possible conformations. The average segment density within a coil, ρ, is low, less than 1 mass %. Noticeable fluctuations in segment concentration take place inside the coil [54, 55]. These fluctuations are on the order of the average polymer concentration in solution and the radius of correlation of these fluctuations is on the order of the size of the coil. Within such a polymer system many-body collisions are rare. The number of pair collisions of segments is large, however. Pair collisions cause the polymer to swell.

The free energy of a polymer consists of the entropic term and the internal energy of the segmental interactions, U, which represents the excluded-volume effect, and can be expanded as a power series of the segment density ρ (Eq. 1):

$$U = VkT\left(\rho^2 A_2 + \rho^3 A_3 + \ldots\right), \tag{1}$$

where V is the volume of the coil, T is the temperature, and A_2 and A_3 are the expansion coefficients. The second virial coefficient, A_2, accounts for binary interactions, whereas the third virial coefficient, A_3, is responsible for three-body interactions. The coefficient A_3 is of importance only in solutions near or below the Θ-temperature. The values of A_2 and A_3 depend on the temperature and on the form of the interaction potential between segments. The second virial coefficient A_2 is defined by the relative distance of a system from the Θ-temperature. It is a measure of the thermodynamic quality of the solvent for the polymer. It can be determined experimentally by light scattering or osmometry.

Under Θ-conditions, i.e. when $T = \Theta$ and $A_2 = 0$, the repeating units of an ideal polymer chain can be described simply as noninteracting molecules of an ideal gas connected by a chain. Under these conditions, the polymer adopts an ideal Gaussian coil conformation. For a flexible high molar mass polymer, the coil size in a Θ-solvent is significantly smaller than its value in a good solvent. In the case of an ideal gas, as the temperature of the system decreases, condensation of the gas takes place. The attractive forces between the molecules become dominant. Similarly, a coil–globule transition of a single polymer chain occurs in a dilute polymer solution when, as the temperature decreases, the solvent becomes thermodynamically poor, $T < \Theta$ and $A_2 < 0$. Condensation of the repeating units of the macromolecular chain in solution takes place. The macromolecule shrinks and becomes more compact than an ideal chain. If there are many chains in the solution, the attraction between the repeating units causes intermolecular aggregation.

Coil and globular conformations are typical examples of the states adopted by a homopolymer in thermodynamically good and poor solvents, respectively. There are some important differences between these two states: the coil conformation corresponds to the lowest minimum of the chain free energy and, therefore, is a thermodynamically stable state. In contrast, the globular conformation of a single chain is a metastable state for the polymer in solution. At the onset of the coil-to-globule transition (Θ-temperature), the polymer–solvent interactions entirely compensate the polymer–polymer interactions: the macromolecule exists in the state of a Gaussian coil, a conformation that corresponds to the global minimum of Gibbs energy and, therefore, is thermodynamically stable. A decrease of the thermodynamic quality of the solvent tends to increase polymer–polymer interactions at the expense of polymer–solvent interactions. Within the transition region (slightly below the Θ-temperature) coils and globules are expected to coexist, depending on the cooling rate and on the kinetics of the transition. Upon cooling, single-chain globules gradually replace coils and the system passes from a stable state to a new metastable state. Globules of single chains tend to associate and adopt a new phase of energy lower than the sum of the energy of individual globules dispersed in the solvent.

It is important to note that, whereas the gas-to-liquid transition is a first-order transition, the introduction of an additional theoretical limitation, namely the connectivity of the repeating units in a long flexible polymer chain, may result in a second-order transition. Also, depending on chain stiffness, the coil–globule transition of macromolecules can be either continuous or discrete. The transition of stiff macromolecules (such as highly charged polyelectrolytes) is an abrupt first-order transition, similar to the gas–liquid transition. However, the coil–globule transition of long flexible chains is a smooth second-order transition, which occurs in a range of temperature broader than in the case of stiff polymers. Flexible macromolecules of infinite molecular weight undergo the transition at the Θ-point, but for real polymers of finite molecular weight, for example polystyrene (PS) dissolved in cyclohexane, the transition occurs below the Θ-temperature [56–58].

The mean-field theory provides a satisfactory description of the experimental results obtained in studies of the coil–globule transition of macromolecules in organic solvents upon abrupt cooling [55, 59–78]. A phenomenological theory suggests a four-stage kinetic process for the polymer collapse induced by a decrease in solution temperature: (1) crumpling of the macromolecule in a single-chain regime, (2) knotting of individual macromolecules and/or rearrangement of thermal beads (also called droplets or pearls), (3) intermolecular aggregation, and (4) interchain entanglement [67, 74, 75, 77]. This theory predicts the existence of a stretched necklacelike conformation as an intermediate stage of the globule formation. The pearls, which are connected by bridges of stretched polymer chains, grow in size and eventually coalesce into a spherical globule. Each transition from a necklace of n pearls to one of $n - 1$ pearls is assumed to be a first-order transition. A phenomenological theory of the necklacelike conformation has been developed by Klushin [79], who introduced a self-similar coarsening stage during the collapse of a flexible macromolecule in a poor solvent. The early stages of the collapse of flexible homopolymers and various phenomenological models of the kinetic process are presented in detail in a recent review by Halperin and Goldbart [77]. Note, however, that only two of the four steps implicated in the theoretical description of the single-chain collapse, the crumpled globule state and the compact spherical globule state followed by aggregation, were observed experimentally in studies of PS in cyclohexane [80–84].

There have been only a few experimental studies aimed directly at monitoring the coil–globule transition of a polymer chain in solution. Indeed, numerous practical problems need to be overcome, especially those associated with intermolecular aggregation and subsequent macroscopic phase separation. To minimise these effects, most experiments have been carried out with macromolecules of high molar mass (above 10^6 g mol^{-1}) and narrow molar mass distribution. To avoid aggregation, measurements have to be done with highly diluted polymer solutions. Scattering methods and fluorescence spectroscopy are best suited to monitoring the coil-to-globule collapse under these circumstances. Light scattering allows one to access quantitative values of the size of a macromolecular coil as it undergoes the transition (hydrodynamic radius, R_h, and radius of gyration, R_g), as well as the thermodynamic parameters associated with the phenomena (the osmotic second virial coefficient, A_2), and the kinetics of the transition. Scattering methods also yield the molar mass of scattering objects, M_w, a quantity required if one wants to study the transition on a molecular level.

For solutions of neutral polymers in organic solvents, the temperature at which phase separation takes place depends on the polymer mass fraction and, in some cases, on the polymer molar mass. The temperature dependence of a polymer–solvent system is conveniently represented in a phase diagram where the phase-separation boundary, or binodal, indicates the temperature for which a given polymer–solvent mixture passes from a one-phase system to a two-phase system consisting of a polymer-rich phase and a polymer-poor phase. In other words, the binodal is the region in the T vs. polymer concentration plot where the coil–globule transition takes place. The maximum of the binodal is the upper critical solution temperature (UCST). Readers are encouraged to consult the recent book by Koningsveld et al. [85], which contains an extensive list of phase diagrams for various binary polymer–solvent mixtures. The book also contains a detailed review of the general thermodynamic principles of phase equilibrium.

PS in cyclohexane is a classic polymer–organic solvent system known to show UCST behaviour: macromolecular collapse upon cooling, followed by macrophase separation [86–88]. Most experiments were conducted on dilute solutions of PS in cyclohexane (less than $c = 10^{-2}$ g L^{-1}) using light scattering [80–84, 86–94] and fluorescence [95, 96]. Baysal and Karasz [97] have recently reviewed the theoretical studies, computer simulations, and experimental studies of the collapse of flexible macromolecules published up to the year 2002. Yamakawa [98], Williams et al. [99], Fujita [100], Des Cloizeaux and Jannink [101], and Klenin [102] contributed earlier reviews on the coil–globule transition. Di Marzio [103] wrote an overview of the various phase transitions in polymeric systems, including the coil–globule transition. Thermodynamic aspects of ultrahigh molecular weight polymers in dilute solutions were summarised by Bercea et al. [104]. Important constants and various empirical relations for real polymers in good and Θ-solvents are given in this review. The role of metastability in the polymer phase transitions over a wide range of polymer concentrations was reviewed by Keller and Cheng [105–107].

In spite of the numerous reported experimental studies of the coil–globule transition, whether or not the true transition has been observed remains subject to discussion [97, 108–110]. The experimental results on the coil–globule transition of PS in cyclohexane have been compared with quantitative results of the mean-field theory by Grosberg and Kuznetsov [73, 74]. Special attention was paid to the stability of PS globules below the Θ-temperature. It was noticed that the precipitation time of PS globules formed by polymers of high molar mass is essentially longer than that of the shorter chains in the solution of the same polymer mass concentration. Grosberg and Kuznetsov also question whether the state of a true, i.e. fully collapsed, polymer chain has ever been observed experimentally owing to intermolecular association. They conclude that at temperatures below the coil–globule transition, the solutions of low enough concentrations cannot be investigated practically with modern instrumentation and sample preparation techniques. The experiments reported deal with solutions at temperatures just below the transition temperature, where one actually cannot distinguish a slightly contracted/crumpled coil from the real globule. Baysal and Karasz [97], as well, doubt that the final, fully collapsed state was actually ever observed. On the basis of their own investigations of the collapse of poly(methyl methacrylate) in isoamyl acetate, they conclude that the total collapse can occur only well below the Θ-temperature and can even be achieved in one step if the depth of quenching (i.e. the shock cooling depth, ΔT) is very deep.

Two facts emerge on turning our attention to aqueous solutions of neutral polymers or polyelectrolytes: these solutions also undergo coil–globule transitions, but the transitions of both neutral water-soluble polymers [80, 83, 84, 111–122] and polyelectrolytes [123–138] differ in many aspects from the coil–globule transition of polymers in organic solvents [86–97]. In order to appreciate this situation, one has to remember that hydrogen bonds, hydrophobic, and hydrophilic interactions contribute much more to the solubility of a polymer in water than van der Waals interactions, which prevail in solutions of polymers in organic solvents. As stated earlier, the experimental value A_2 reflects the balance of these various interactions. The collapse of a homopolymer upon decreasing the thermodynamic quality of the solvent can be classified as a coil–globule transition if $A_2 < 0$ in the collapsed state. However, water-soluble polymers in the collapsed state in water may adopt conformations other than the "classic" globular state of PS in cyclohexane, for which $R_g/R_h \sim 0.78$ as for a hard nondraining sphere. This condition is not always fulfilled in the case of aqueous polymer solutions [139, 140].

Phase transitions of aqueous polymer solutions have been the subject of several earlier reviews [97–104]. In this review, we will focus primarily on the collapse of neutral water-soluble polymers in aqueous media and the various physicochemical methods that have been used to enhance the colloidal stability of polymeric particles formed under thermodynamically poor conditions. As stated in several recent publications on the thermal behaviour of polymeric aqueous solutions [141–147], prior results may need to be reassessed, given the fact that early experiments were often performed in the presence of small amounts of external stabilising agents that may have contributed inadvertently to the stability of polymeric particles in their collapsed state.

In analogy with solutions of neutral polymers in organic media, solutions of nonionic water-soluble polymers in water can be described by phase diagrams with a UCST, a lower critical solution temperature (LCST), and/or miscibility loops [34]. Many of these systems have an LCST which reflects a local structural transition involving water molecules surrounding specific segments of the polymer in solution. At low temperatures, water molecules are "frozen" in place via hydrogen bonds with functional groups of the polymer and among themselves. They provide the polymer with a hydration layer that effectively renders it soluble in water. At higher temperatures, these molecules are released or "melt", allowing associative contacts between the newly exposed monomer units [38]. This may lead to chain contraction, and eventually to phase separation. In other cases, it may lead to gelation [148]. Among the first systematic studies of aqueous thermoresponsive polymers, the 1975 work of two industrial scientists, Taylor and Cerankowski [149], stands out. On the basis of a set of cloud-point values gathered from solutions of a broad range of synthetic polymers, the authors state that by substituting hydrophilic groups that make a polymer water-soluble with hydrophobic groups, one can convert a polymer, originally soluble in water at all temperatures, into a polymer soluble in water only below a given temperature.

The solution properties of three thermosensitive polymers, poly(*N*-isopropylacrylamide) (PNIPAM) [150], poly(*N*-vinylcaprolactam) (PVCL) [151, 152], and poly(vinyl methyl ether) (PVME) [153–156], have been scrutinised in great depth. Aqueous solutions of PNIPAM have an LCST of approximately 31 °C, regardless of the molar mass of the polymer [150], except in the case of oligomeric PNIPAMs for which higher LCST values have been reported. The cloud point of aqueous PVCL solutions lies above 30 °C and significantly increases with decreasing PVCL molar mass [151]. Aqueous solutions of PVME exhibit two LCSTs at temperatures of approximately 32 and approximately 40 °C. The values depend on both the polymer molar mass and on its concentration [153, 154]. It seems therefore that there may not be a universal mechanism applicable to the phase transition of all polymers in water. To facilitate the description of the phase-separation phenomenon of aqueous polymer solutions, it is useful to classify them according to the phenomenological analysis of their critical miscibility with water. Polymers of type I follow the classic Flory–Huggins behaviour, that means that the LCST (i.e. the minimum of the phase diagram) of polymers of this type shifts upon increasing the polymer molar mass towards lower polymer concentrations (Sect. 3.1; Fig. 14). For type II polymers, the chain length does not influence the position of the minimum in the demixing curves (Sect. 2.1; Fig. 3). Phase diagrams of type III polymers are bimodal, presenting two critical points at low and at high polymer concentrations corresponding to type I and type II behaviours, respectively (Sect. 4; Fig. 31). According to this classification, PVCL is a type I polymer [157, 158], PNIPAM is an example of type II polymers [159, 160], and PVME is a type III polymer [161–164]. This phenomenological classification, although somewhat arbitrary, has been adopted in this review, with Sects. 2, 3, and 4 dedicated to polymers of types II, I, and III, respectively.

The increase in the hydrophobicity of neutral polymers in their collapsed state at elevated temperatures causes aggregation, unless special care is taken to prevent interpolymer chain association. Thus, to avoid, or at least to minimise, intermolecular contacts, it is best to work with dilute solutions. Another strategy is to strive towards colloidal stability by linking to the polymer chain substituents, which preferably reside near the water interface of the collapsed globules, rendering the interface nonadhesive or repulsive towards other globules. The latter approach is challenging since, as stated by Taylor and Cerankowski [149], the coil–globule transition can easily be shifted to higher or lower temperatures by increasing the fraction of hydrophilic or hydrophobic groups, respectively. Kinetic and thermodynamic control of the thermosensitivity is essential and careful adjustments in the polymer composition and topology are needed.

The colloidal stability of intermolecular aggregates formed by thermally responsive polymers under thermodynamically poor conditions can be achieved by modifying the surface of the particles either sterically [165–172] or electrostatically [114, 115, 141–145, 173–176]. For instance, stabilisation of single-chain PNIPAM globules was achieved by addition of ionic surfactants by Ricka et al. [114, 115], who used small concentrations of sodium dodecyl sulphate (SDS) to prevent aggregation. They demonstrated that the coil–globule transition of PNIPAM occurs even in the presence of SDS and that the repulsion between the charged head groups of the SDS molecules bound to the polymer chains stabilises the polymeric globule. However, SDS alters the LCST value of aqueous PNIPAM solutions, which complicates data interpretation: the structure of polymer–SDS complexes may differ from that of collapsed globules and the distribution of SDS in solution and bound to the polymer was not assessed. Introduction of positively [141, 173] or negatively [176] charged comonomers also leads to the formation of stable aggregates at temperatures higher than the LCST. Steric stabilisation of the collapsed globules has been achieved by decorating the surface of the aggregates with neutral hydrophilic substances, such as PEO, as discussed in detail in Sects. 2.2 and 3.2. Thus, for PNIPAM grafted with PEO, sterically stabilised spherical particles form above the LCST [165–168]. PVCL with amphiphilic PEO grafts show similar properties [146, 177–182]. Copolymerisation of *N*-isopropylacrylamide (NIPAM) with hydrophilic vinylpyrrolidone (VP) at temperatures above and below its LCST results in segmented and random distribution of VP units, respectively [183]. The copolymers with a segmented VP distribution aggregate more readily upon heating and form stable aggregates larger than those with a random VP distribution. This leads one to conclude that not only the presence of hydrophilic groups, but also their distribution along the polymer chain play an important role in the stabilisation of globular particles under thermodynamically poor conditions.

On the basis of this controllable change of conformation, and consequently solubility, of a polymer in water, various "smart" water-borne systems may be imparted with a specific response to a sudden or progressive change in temperature. Moreover, thermally responsive polymers can be cross-linked to form responsive hydrogels. Porous surfaces, grafted with smart polymers, exhibit temperature-dependant porosity and may be used in thermally responsive filter units [184]. Numerous other applications are under current investigation in areas such as controlled release, drug delivery, bioseparation, and diagnostics.

1.4 Scope of the Present Review

In the previous sections, we described the overall features of the heat-induced phase transition of neutral polymers in water and placed the phenomenon within the context of the general understanding of the temperature dependence of polymer solutions. We emphasised one of the characteristic features of thermally responsive polymers in water, namely their increased hydrophobicity at elevated temperature, which can, in turn, cause coagulation and macroscopic phase separation. We noted also, that in order to circumvent this macroscopic event, polymer chemists have devised a number of routes to enhance the colloidal stability of neutral globules at elevated temperature by adjusting the properties of the particle–water interface.

It turns out that in solutions of $c < 0.1\,\mathrm{g\,L^{-1}}$ thermosensitive homopolymers, such as PNIPAM, PVCL, and PVME, themselves, form stable colloids in water at elevated temperature in the absence of additives or chemical modification [141–147]. The colloids remain stable upon prolonged heat treatment, without detectable aggregation or precipitation. Also, core–shell particles consisting of PNIPAM and a hydrophobic block are stable not only below but also above the LCST up to 50 °C, when the PNIPAM block is expected to be insoluble [185]. Factors that determine the colloidal stability as defined in Sect. 1.1 do not explain, it seems, their stability. In this review we have compiled a list of all the reported instances where the formation of stable particles was detected in aqueous solutions of neutral thermosensitive neutral polymers at elevated temperature. We present studies of homopolymers, as well as their copolymers consisting of thermosensitive fragments and either hydrophilic of hydrophobic fragments. We will also present the various explanations put forward to account for the stability of these particles.

2 Poly(*N*-alkylacrylamides)

2.1 Poly(*N*-isopropylacrylamide)

The temperature-dependent solubility of PNIPAM (Fig. 1) in water was reported for the first time in 1963 in a technical brochure on NIPAM and its polymer, where potential users of the polymer were made aware of the fact that the solubility of PNIPAM in water gradually decreases with increasing temperature [186]. Scarpa et al. [187] reported the sharpness of the transition in 1967. The first detailed study of aqueous PNIPAM solutions as a function of temperature was reported in 1969 by Heskins and Guillet [188]. The authors observed visually the change in clarity of a solution upon heating and determined that macroscopic phase separation of PNIPAM solutions takes place at 32 °C. Since then numerous publications on PNIPAM, its derivatives, and its applications have appeared. The reader is referred to a review on PNIPAM [150], written in 1992, for a general, although slightly dated, overview of the properties of PNIPAM, as well as publications on specific aspects of PNIPAM solutions and microgels [189–198]. We will limit this review to articles focussed on the thermal properties of PNIPAM and its derivatives, as they pertain to our primary concern, the colloidal stability of PNIPAM globules at elevated temperature.

Turbidimetry is ideally suited to detect the temperature at which a transparent polymer solution turns opaque. The temperature corresponding to the onset of the increase of the scattered light intensity is usually taken as the cloud-point temperature, T_{cp}, although some authors define the cloud point as the temperature for which the transmittance is 80% (or 90%) of the initial value. This technique is commonly known as the cloud-point method [199]. Turbidimetry was employed, for instance, to show that the cloud-point temperature of aqueous PNIPAM solutions does not depend significantly on the molar mass of the polymer [150].

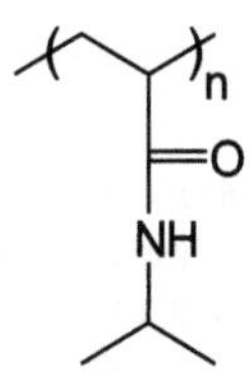

Fig. 1 Chemical structure of poly(*N*-isopropylacrylamide) (*PNIPAM*)

Differential scanning calorimetry (DSC), in particular high-sensitivity DSC (HS-DSC), provides another look at the heat-induced phase transition that takes place in a polymer solution. This technique allows one to determine the thermodynamic parameters of the phase transition: the temperature of maximum heat capacity, T_{max}, the width of the transition at half-height, $\Delta T_{1/2}$, the heat of transition, ΔH, and the difference in the heat capacity before and after the transition (ΔC_p); see Fig. 2. Afroze et al. [159] reported recently a study of the demixing of water and PNIPAM of varying chain lengths over a wide range of composition. They defined the demixing temperature, T_{dem}, as the temperature corresponding to the onset of the endothermic signal in the DSC trace (Fig. 2). They also determined the cloud points of the mixed systems by turbidimetry (Fig. 3a). Values of T_{cp} and T_{dem} agree reasonably well. The small discrepancy between the two values for a given water–PNIPAM mixture and a given polymer molar mass may be ascribed to differences in the heating rates employed in the two methods. The authors point out that the optical method yields slightly different cloud points ($T_{cp} < T_{dem}$) since it is less prone to supercooling or superheating than the calorimetric technique [85]. The phase diagram presents a minimum with a broad miscibility gap for of PNIPAM–water weight fractions around 0.4–0.5 (Fig. 3a). The minimum in the demixing curves is not affected by PNIPAM molar mass or, stated otherwise, PNIPAM exhibits a type II demixing behaviour. Note that both T_{cp} and T_{dem} increase for solutions of polymer weight fractions lower than 0.4, the effect being more pronounced as the chain length decreases.

A number of investigations on the thermal volume transition of PNIPAM have been performed [201–208]. PNIPAM chains carry two types of bound

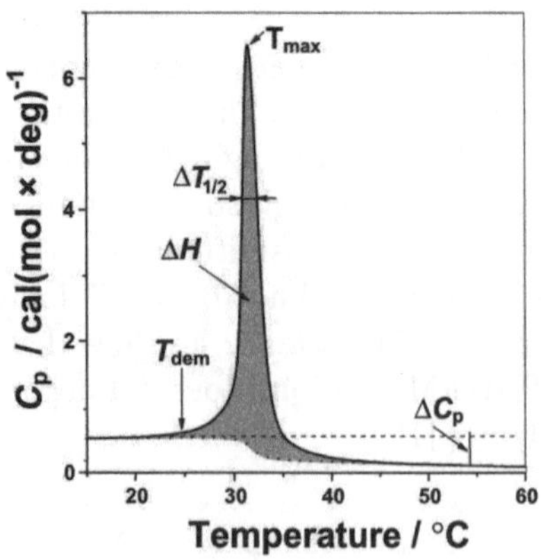

Fig. 2 Typical thermogram obtained using conventional differential scanning calorimetry on PNIPAM solution: the temperature of maximum heat capacity (T_{max}), the width of the transition at half-height ($\Delta T_{1/2}$), the heat of transition (ΔH), the difference in the heat capacity before and after the transition (ΔC_p), and the demixing temperature (T_{dem}). (Adapted from Ref. [200])

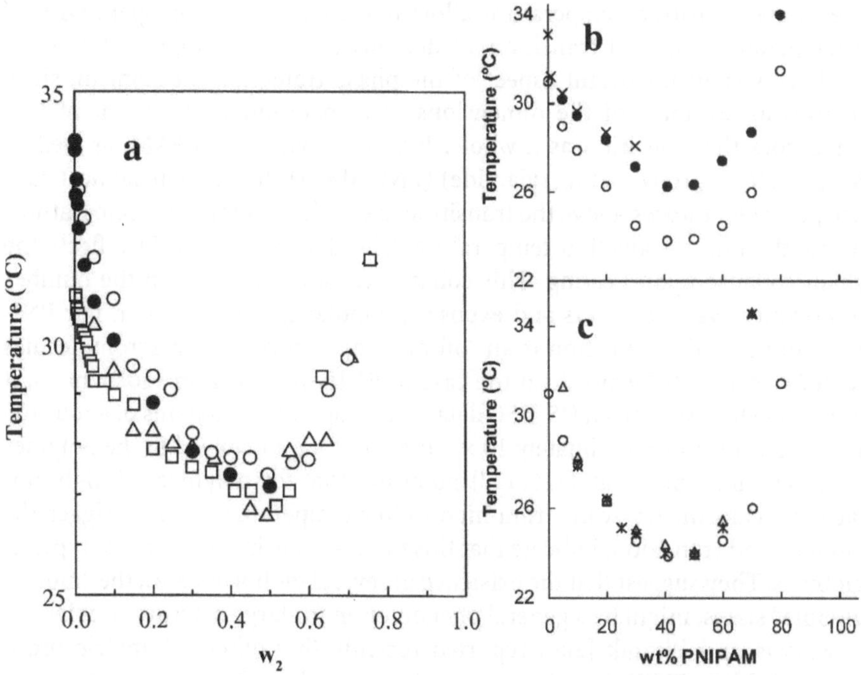

Fig. 3 a Type II phase diagram of for aqueous PNIPAM of three different molar masses: T_{dem} of $M_w = 124\,000\,g\,mol^{-1}$ (□), T_{dem} of $M_w = 53\,000\,g\,mol^{-1}$ (△), T_{dem} of $M_w = 10\,000\,g\,mol^{-1}$ (○), cloud-point temperature (T_{cp}) of $M_w = 10\,000\,g\,mol^{-1}$ (•). **b** PNIPAM 74 000–water, demixing curve: threshold in C_p $0.01\,J\,g^{-1}\,K^{-1}$ (○), $0.1\,J\,g^{-1}\,K^{-1}$ (•) and cloud-point curve (×). **c** Demixing curves for PNIPAM 18 000–water (△), PNIPAM 74 000–water (○) and PNIPAM 186 800–water (∗). Demixing temperatures are calculated from the threshold in C_p (0.01 or $0.1\,J\,g^{-1}\,K^{-1}$) upon heating. (Reprinted with permission from Ref. [159] copyright 2000 Elsevier and Ref. [160] copyright 2004 American Chemical Society)

water molecules in the so-called water cages around the isopropyl groups. One of them is the water cage around the hydrophobic moiety; the other one is the bound water around the amide group. It has been suggested that the transition of PNIPAM consists of at least two different processes: a rearrangement of bound water around either the hydrophobic moieties only or around both hydrophobic and hydrophilic groups. The change in the hydration state of the hydrophobic side chains results in hydrophobic association of the hydrophobic groups and in a residual interaction between the side chain residues [201]. However, the phase transition may also be the result of dissociation of water molecules around the hydrophobic PNIPAM groups [205, 206, 208].

Data gathered by microcalorimetry, specifically changes in the partial heat capacity, $C_p(T)$, of a polymer solution at temperatures below and above the phase-transition temperature allow one to assess the cooperativity of a temperature-induced transition, as demonstrated by Tiktopulo et al. [116, 117]. This is an important aspect of the phase-transition mechanism, since it gives an estimate of the dimensions of a macromolecule fragment that undergoes the transition as a whole. It was shown for PNIPAM, as well as for poly(*N*-isopropylmethacrylamide) (PNIPMAM) that the partial heat capacity at temperatures above the transition is smaller than that at temperatures below the phase-transition temperature. The decrease in $C_P(T)$ reflects the chain collapse upon heating. This collapse causes a decrease in the number of polymer–water contacts and exposes nonpolar groups to water. For PNIPAM in water the transition is an "all-or-none" process for macromolecules with $M_W < 10 \times 10^3\ \mathrm{g\,mol^{-1}}$. In the case of PNIPAM of higher molar mass, it is possible to assess, from HS-DSC data, that cooperative domains of about 100 monomer units (approximately $10 \times 10^3\ \mathrm{g\,mol^{-1}}$) are involved in the polymer collapse. Tiktopulo et al. [116, 117] point out that for polymers of high molar mass calorimetry scans from high to low temperature, which trigger the globule–coil transition, indicate that this process may involve various types of globules. They suggest that the existence of several globular states, the "molten globule" states, might be a general phenomenon for large macromolecules.

Kujawa and Winnik [209] reported recently that other volumetric properties of dilute PNIPAM solutions can be derived easily from pressure perturbation calorimetry (PPC), a technique that measures the heat absorbed or released by a solution owing to a sudden pressure change at constant temperature. This heat can be used to calculate the coefficient of thermal expansion of the solute and its temperature dependence. These data can be exploited to obtain the changes in the volume of the solvation layer around a polymer chain before and after a phase transition [210], as discussed in more detail in the case of PVCL in Sect. 3.2.2.

Application of temperature-modulated DSC (TM-DSC) allows one to assess the relative importance of the bound water–polymer interactions and the hydrophobic interactions during the phase transition of PNIPAM aqueous solutions, and gives access to the kinetics of the demixing/remixing processes [160, 211]. Details of the method can be found elsewhere [212, 213]. A single endotherm was separated into endothermic and exothermic contributions, of variable intensity, depending on the frequency applied. Curve-fitting to experimental data suggests that the most realistic model for the transition mechanism of aqueous PNIPAM involves three processes. This implies that the phase transition does not seem to occur only by restructuring the water molecules around the hydrophobic group in the PNIPAM chain. Instead, in the three-processes model, the water molecules around the hydrophilic group also rearrange to bulk water, although it is not known which portion of these water molecules participates in this process. It is also thought that the intermolecular interaction between PNIPAM chains is not attributable only to hydrophobic association of the hydrophobic groups. Some of the intermolecular hydrogen bonds between the amide groups can also contribute to the heat of transition with hydrogen-bond formation between sites vacated by water molecules displaced from the amide groups.

A large number of experimental studies on the coil–globule transition of PNIPAM in water have been done on very dilute solutions of high molecular weight PNIPAM ($M_w \sim 10^7\,\mathrm{g\,mol^{-1}}$) using light scattering methods [111–122, 193, 194]. Through a combination of static light scattering and dynamic light scattering (DLS) measurements, the ratio of the mean-square radius of gyration, R_g, to the hydrodynamic radius, R_h, of a PNIPAM chain was determined in solutions kept at temperatures below and above the phase-transition temperature. The ratio R_g/R_h takes a value around 0.78 above the critical temperature and around 1.5 below this temperature, values corresponding to the theoretical values for hard uniform spheres and monodisperse Gaussian random coils, respectively [214]. From the values of R_g/R_h and the average density, ρ, obtained at elevated temperature, one can conclude that a PNIPAM chain in its globular state still contains about 70–80% of water [84, 119]. A comparison of light scattering data monitoring the coil–globule and the globule–coil transitions of aqueous PNIPAM indicates that the two processes do not follow the same mechanistic path, a conclusion also reached on the basis of calorimetric data (see before). During the coil–globule–coil transition, a PNIPAM chain is believed to adopt four different consecutive states: coil, folded coil, fully collapsed globule, molten globule, and coil [119, 121, 122]. Two globular states of PNIPAM (crumpled and fully collapsed globule) have also been observed by Napper [118] using light scattering.

Kinetic aspects of the coil-to-globule collapse of a PNIPAM chain were investigated by Yu et al. [80] and Chu et al. [83, 84], who performed light scattering measurements on PNIPAM solutions heated at a very slow rate, i.e. step-by-step heating. This allowed them to study the coil–globule transition under equilibrium conditions [84]. The data were compared with those obtained by fast, quench overcooling of a PS solution in cyclohexane (from 35 to 28 °C in approximately 100 s) [80, 83]. Both PS and PNIPAM systems pass from a homogeneous one-phase solution to a metastable two-phase system in a similar manner. However, in the case of PS in cyclohexane, the collapse can be achieved kinetically by quickly quenching the solution below the coexistence curve, which separates the one-phase region from the two-phase region, to avoid aggregation of globules [84]. The aggregation of single-chain PS globules starts after about 10 min [83]. In the course of a light scattering study, Wu and Zhou [120] made a key observation, at least in the context of this review: they noted that the size of the PNIPAM globules formed above the phase-separation boundary remained unchanged when samples were kept at a temperature higher than the phase-transition temperature for a period of 3 days.

The reluctance of single PNIPAM globules to aggregate at elevated temperature may be ascribed qualitatively to the low probability of encounter of two globules under the high dilution conditions employed. However, several investigators noted the stability against macroscopic aggregation/precipitation of clusters or multimolecular aggregates in much more concentrated PNIPAM solutions [141–147]. For example, Chan et al. [141] reported the formation of colloidally stable dispersions of microphase-separated PNIPAM ($M_W = 547\,000\,\mathrm{g\,mol^{-1}}$) in solutions of concentration up to $1.0\,\mathrm{g\,L^{-1}}$. The diameter of the PNIPAM aggregates increased as a function of the cube root of PNIPAM concentration. The following explanation was offered to account for the unexpected stability of aggregates. The specific PNIPAM sample investigated was obtained by persulphate-initiated free-radical polymerisation of NIPAM, and, the authors concluded this type of polymerisation is known to introduce a small number of negative charges on the polymer chain. Consequently, the stability of the collapsed PNIPAM globules may stem from the residual charges and may be a typical case of electrostatic colloidal stability. Under the conditions employed, individual globules are not colloidally stable; they coagulate, yielding larger particles. These grow in size until a colloidally stable population of particles is reached. Thus, the aggregation is entropically unfavourable and the van der Waals forces of attraction between particles increase with particle size. To explain the experimental phenomenon, Chan et al. assumed that the electrostatic energy of repulsion increases faster than the van der Waals attraction. They monitored the effect of salt addition on particle stability and related the critical diameter of aggregates to their colloidal stability. They pointed out that the single-chain globules reported by Wang et al. [121] in their study of dilute PNIPAM of $M_W = 1.3 \times 10^7\,\mathrm{g\,mol^{-1}}$ contain about 66% water. The estimated half-life of such a dispersion is of the order of 4 s [141], a value different from the experimental observation [120, 121], suggesting that the stability of single-chain PNIPAM globules may not be due only to the high dilution conditions of the experiments.

The formation of mesoscopic aggregates of PNIPAM [$M_W = (0.5-10) \times 10^6\,\mathrm{g\,mol^{-1}}$] in solutions above the phase-separation boundary was also reported by Gorelov et al. [144, 145], who employed PNIPAM samples obtained via using persulphate-initiated polymerisation of NIPAM as well [141]. They carried out a light scattering study of aqueous PNIPAM ($c < 0.2\,\mathrm{g\,L^{-1}}$) heated up to 65 °C. In solutions heated above the phase-transition temperature, stable particles formed. Their size depended on polymer concentration, increasing with increasing initial concentration, and on heating rate, decreasing with increasing heating rate. In all cases the particle size distribution was narrow [144]. Gorelov et al. found that for solutions kept well above the transition temperature (35–65 °C), the size of the particles remained constant over periods of several days and was stable upon dilution.

The following mechanism emerges form the observation reported so far on the colloidal stability of multichain PNIPAM particles at elevated temperature. Let us note first that such PNIPAM particles are examples of mesoglobules, according to the definition of Timoshenko and Kuznetsov [215]:

mesoglobules are "equally sized spherical aggregates of more than one and less than all polymer chains colloidally stable in solution". Two factors may contribute to the stability of PNIPAM mesoglobules: they are related to the spinodal decomposition of polymer solutions above the LCST and to the low probability of Brownian collisions in dilute solutions [145]. Thus, a polymer in a poor solvent optimises its interaction energy by intermolecular association among polymer chains, which results in a decrease of the translational entropy of the chains. When these two energy terms are compensated, the system reaches a metastable state characterised by the formation of polymeric mesoglobules. Another and possibly more important cause of the mesoglobule stability might be that the aggregates are quite dense and compact, and, consequently, the merging of two mesoglobules into a single, larger one may take a long time. Such a state appears as a transient nonequilibrium state after a quench to the poor-solvent region. In other words, the metastable state is kinetically arrested, since the coalescence time of solidlike particles is large, compared with the characteristic collision time. The theory of mesoglobule formation has been developed and applied for block and random copolymers [145, 216, 217]. It should be pointed out at this point that other factors might contribute inadvertently to the stability of PNIPAM mesoglobules. For instance, it has been reported that self-cross-linking of growing PNIPAM chains takes place in the absence of any cross-linking agent, when PNIPAM is prepared by precipitation in water at elevated temperatures [218–220].

In this section, we made an inventory of situations where the thermosensitive homopolymer PNIPAM forms mesoglobules stable against aggregation at elevated temperature for extended periods of time. We noted that stable single-chain globules exist at elevated temperature under conditions of high dilution, and that multichain mesoglobules of remarkable stability towards macrophase separation form under controlled conditions and within certain concentration limits. Several hypotheses have emerged to account for the stabilisation mechanism at work, since the typical electrostatic or steric mechanisms underlying classic colloidal stability do not seem applicable. We present next the various methods employed to prepare PNIPAM mesoglobules/particles specifically designed to remain stable against aggregation at elevated temperature via either steric or electrostatic interactions. They involve chemical modification of PNIPAM via introduction of amphiphilic fragments, mostly PEO, either grafted along the PNIPAM chain, PNIPAM-*g*-PEO, or linked to one chain end, PNIPAM-*b*-PEO.

2.2 Polymeric Self-Assemblies with a Core–Shell Structure

In dilute aqueous solutions, copolymers having hydrophobic and hydrophilic parts may form polymeric micelles, i.e. stable particles with a core–shell structure. The association of the hydrophobic parts of the block copolymers occurs at a given polymer concentration, often regarded as a critical micelle or aggregation concentration which greatly depends on the hydrophilic–lipophilic balance of the polymer. In aqueous solutions of stimulus-responsive, intelligent copolymers, the interaction of one or several parts of the copolymer can be triggered by changes in the surroundings of the spherical core–shell structures [221–227]. They are switchable amphiphiles in which the solubility properties of each block can be altered by changing the solution pH [228, 229], ionic strength [229], or temperature [165–172, 177–182] or by interaction with added substrates [230–233].

There are numerous examples on the use PEO chains to induce stability against aggregation of particles, such as polymeric micelles, liposomes, and metal particles. The efficiency of steric stabilisation is based on the surface coverage of the hydrophilic PEO chains on the particle. For example, polyanions and polycations have been synthesised that either have a PEO block in the chain end or PEO grafts along the chain [234–239]. Mixing a pair of oppositely charged polyions results, at the point of charge neutralisation, in the formation of hydrophobic complexes. Water turns out to be a poor solvent for such complexes. They associate, decreasing the overall entropy of the solution. Upon mixing of equimolar amounts of PEO–polyanion and PEO–polycation diblocks in water, precipitation does not occur, instead stable particles form, the PEO chains providing steric stabilisation to the neutralised, hydrophobic polyanion–polycation complex. The formation of polyelectrolyte complexes is a complicated process and the structures of the particles are affected by numerous factors [240–243]. Very often, the particles may be assumed to be electrostatically stabilised owing to uncompensated charges and, thus, these systems are not discussed further in the present review.

The surface coverage and the conformation of PEO chains on spherical hydrophobic polymer particles have been studied extensively [244]. Briefly, PEO chains lie flat on the surface of a particle when the number of PEO chains is low. With increasing number of PEO chains the packing density on the particle surface increases and the PEO chains extend far into the continuous medium. This change has been termed a pancake–brush transition [245, 246]. This model is also relevant in the context of this review, with regard to the aggregate formation of block and graft copolymers of NIPAM and ethylene oxide (EO). Above the temperature of the PNIPAM coil–globule transition, the copolymers form spherical particles, sterically stabilised by an outer shell of PEO. In these particles PEO chains are stretched out towards the aqueous phase as depicted schematically in Fig. 4.

The formation of stable nanoparticles has been studied using various derivatives of thermosensitive PNIPAM, including diblock and graft copolymers, PNIPAM-*b*-PEO and PNIPAM-*g*-PEO [165–172]. In these copolymers, the role of the PEO chains is to solubilise/stabilise collapsed PNIPAM at temperatures above its cloud point. Both the graft and the block copolymers, PNIPAM-*g*-PEO and PNIPAM-*b*-PEO, form spherical core–shell structures in

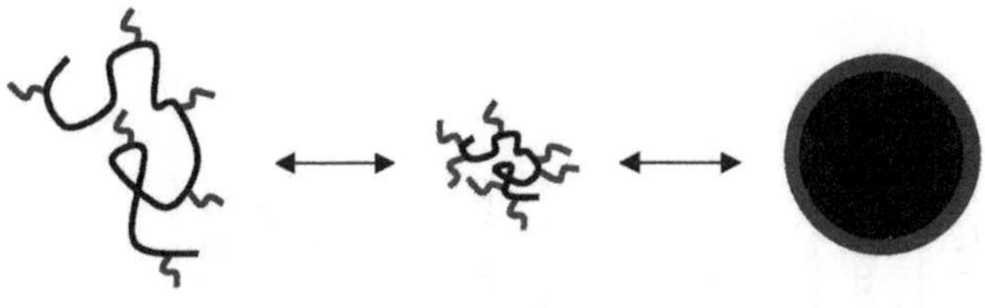

Fig. 4 Formation of a core–shell sphere upon heating an aqueous PNIPAM-*g*-poly(ethylene oxide) (*PEO*) copolymer solution: *black* PNIPAM backbone and *grey* PEO grafts. *LCST* lower critical solution temperature [170]

dilute aqueous solutions at or above a critical temperature. Factors determining the coil–globule transition of the thermosensitive part include intrachain and interchain hydrophobic interactions, and the solubilising effect of the hydrophilic shell on the shrunken backbone.

2.2.1
PNIPAM-*g*-PEO Copolymers

This section summarises the thermal properties of aqueous solutions of various NIPAM copolymers, focusing on the effect of the number and spatial distribution of PEO grafts on the shrinking and collapse of the PNIPAM main chain during heating and on the overall stability of the resulting particles at elevated temperature. Several research groups have prepared PNIPAM-*g*-PEO and examined the solution properties and applications [247–253]. Worth noting is the work of Qiu and Wu [247], who synthesised a PNIPAM-*g*-PEO graft copolymer with high molar mass, and studied the coil–globule transition of a single chain. The methodology exploited by Virtanen et al. [165–168, 170] will be described in detail in the following sections, as the studies were designed to address in a systematic way the factors affecting the colloidal stability of PNIPAM-*g*-PEO.

PNIPAM-*g*-PEO graft copolymers were prepared in two steps, shown schematically in Fig. 5 [165–168, 170]. In the first step, NIPAM copolymers carrying reactive units were synthesised by free-radical copolymerisation of NIPAM and either glycidylmethacrylate (GMA) or *N*-acryloylsuccinimide (NASI). The comonomers GMA and NASI were chosen as functional groups owing to their high reactivity towards primary amine groups [254–256] (Table 1). Kinetic studies of the copolymerisation of NIPAM and GMA showed the latter monomer to be more reactive than the former. In order to obtain a random distribution of the functional groups along the PNIPAM chain, GMA was added continuously to the polymerisation mixture.

In the second step, amino-terminated PEO chains were linked to the reactive NIPAM copolymers by condensation reaction of the primary amine

Table 1 Characteristics of graft copolymers of poly(*N*-isopropylacrylamide) (*PNIPAM*) and poly(ethylene oxide) (*PEO*) [165–167, 170]

Sample	Comonomer or NH_2-PEO (mol %) In feed	In polymer[a]	Number of PEO grafts[b]	PEO (wt %)	$M_w \times 10^{-5}$ ($g\,mol^{-1}$)	Grafting solvent (temperature)
PNIPAM-*co*-GMA	5.0	1.5			1.80[c]	
PNIPAM-*g*-PEO-6	2.3	0.38	6	17	2.16[d]	Water (15 °C)
PNIPAM-*g*-PEO-7	2.3	0.44	7	19	2.22[d]	Water (29 °C)
PNIPAM-*g*-PEO-10	3.0	0.6	10	25	2.40[d]	Dioxane (reflux)
PNIPAM-*co*-NASI	2.5	4.0			1.90[c]	
PNIPAM-*g*-PEO-43	5.3	2.5	43	57	4.48[d]	Dioxane (reflux)
PNIPAM-*g*-PEO-51	5.3	3.0	51	61	4.96[d]	Dioxane (reflux)
PNIPAM-*g*-PEO-57	5.3	3.3	57	65	5.32[d]	Dioxane (reflux)
PNIPAM-*g*-PEO-79	5.3	4.0	79	72	6.64[d]	Dioxane (reflux)

[a] Determined by ^{1}H-NMR or ^{13}C-NMR
[b] Determined by ^{13}C-NMR
[c] Measured by static laser light scattering
[d] Calculated by M_w (backbone polymer) + number of PEO grafts $\times M_w$ (PEO)

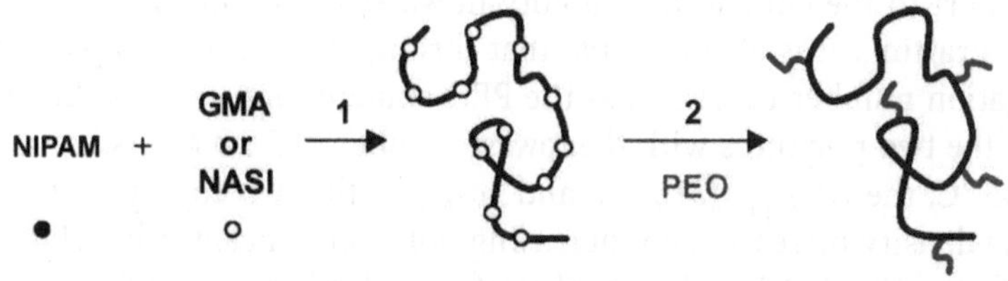

Fig. 5 Synthesis of the graft copolymers in two steps: (*1*) copolymerisation of *N*-isopropylacrylamide (NIPAM) and glycidylmethacrylate (*GMA*) or *N*-acryloylsuccinimide (*NASI*), and (*2*) grafting of PEO to the polymer backbone [170]

group with the polymer-linked functional groups. In the case of NIPAM-GMA ($M_w = 180\,000\,g\,mol^{-1}$), grafting reactions were carried out (1) in water at 15 °C, (2) in water at 29 °C, and (3) in refluxing dioxane. In all cases, the same amine-terminated PEO ($M_w = 6000\,g\,mol^{-1}$) was employed and all attempts were made to achieve the same grafting density. This was an important consideration, since one of the objectives of the study was to assess the effect of grafting conditions on the solution properties of the copolymers. The copolymer NIPAM-NASI was treated with increasing amounts of PEO-NH_2 in dioxane, with the objective of creating a series of copolymers of varying degree of grafting obtained under the same conditions (Table 1).

The phase-separation process was monitored using light scattering. Aqueous solutions of the reactive copolymers PNIPAM-*co*-GMA and PNIPAM-*co*-NASI ($c = 1.0\,g\,L^{-1}$) exhibited cloud points at 31.0 ± 0.1 and 31.8 ± 0.1 °C, respectively, values close to the cloud point of PNIPAM itself. Introduction of PEO grafts along PNIPAM had a marked effect on the phase-transition temperature, which increased from 32.5 to 34.7 depending on the number of PEO grafts per PNIPAM chain (Table 1). This increase reflects the competition between the hydrophobic interactions and the solubilising effect of the hydrophilic PEO chains.

In all cases, the cloud-point temperature was slightly dependent on polymer concentration: for a given copolymer it increased with decreasing concentration. This effect is enhanced with increasing number of PEO grafts per chain. Also, the PNIPAM collapse seemed to be less abrupt with decreasing concentration. Upon dilution of the solution the distance between polymer chains increases, which favours intrapolymeric interactions over interpolymeric attractions. Dilution also enhances the surface stabilisation of the polymer particles by PEO.

Raising the temperature of aqueous PNIPAM-*g*-PEO-51 aqueous solutions from 20 to 60 °C resulted in a sudden increase in scattering intensity at 32–33 °C (Fig. 6). The intensity of the light scattered by the aggregates above the cloud point decreased with increasing PEO content, indicating that the average density of the particles decreased. This may be due in part to the ability of the PEO grafts to prevent partially the collapse of the PNIPAM chains and in part to the enhanced solubility in water of polymers with higher degree of grafting. It is also possible that through the whole sample series the aggregation number decreases as the PEO content increases, as shown in the case of the two polymers with the lowest number of PEO grafts.

At 45 °C, the $\langle R_h \rangle_{agg}$, $\langle M_w \rangle_{agg}$, and $\langle \rho \rangle_{agg}$ of the two copolymers of lowest grafting density increase with increasing polymer concentration (Fig. 7). The shape factor $\langle R_g \rangle / \langle R_h \rangle$ is close to that of a hard sphere, 0.778 for the solution of highest polymer concentration (0.8 g L^{-1}). As the polymer concentration decreases, the ratio $\langle R_g \rangle / \langle R_h \rangle$ decreases, reaching a value lower than that of a hard sphere. As mentioned earlier, Wu and Zhou obtained a value of 0.62 in their study of single PNIPAM chains in extremely dilute solution (approximately 5 μg L^{-1}). They ascribed this unusually low value to the morphology of the PNIPAM mesoglobules having a higher chain density in the centre, compared with the periphery [119]. This interpretation may also apply in the case of PNIPAM-*g*-PEO discussed here.

Some aggregated polymer solutions were diluted with water, keeping the solutions at 45 °C. The diluted samples were equilibrated for 24 h, and $\langle R_h \rangle_{agg}$ was measured again. The sizes of the aggregates decreased only slightly upon dilution, indicating that the aggregates, once formed, do not easily reorganise. This implies that polymer chains in the core of the aggregates are kinetically trapped.

Recent theoretical studies suggest that copolymers consisting of structural units with different solubilities may remember the conformations they adopted during their synthesis [257–260]. Such polymers, of which PNIPAM-

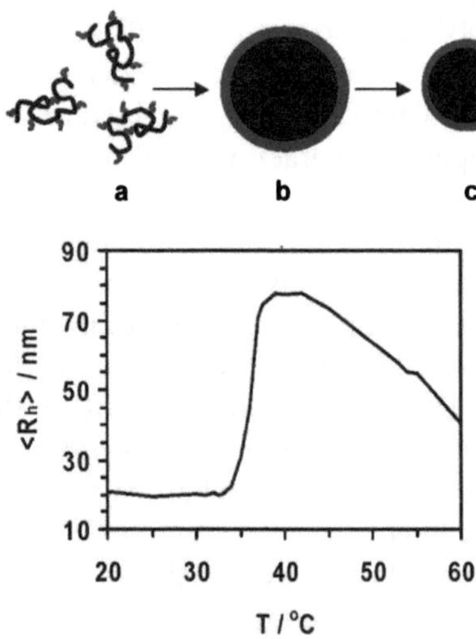

Fig. 6 Formation of an aggregate and the dependence of its average hydrodynamic radius ($\langle R_h \rangle$) on temperature. The polymer is PNIPAM-*g*-PEO-51, $c = 1.0$ g L^{-1}. *Top*: model describing the steps for the formation of an aggregate and its shrinking upon slow heating from *a* 20 °C to *b* 45 °C and to *c* 60 °C. (Adapted from Refs. [165, 170])

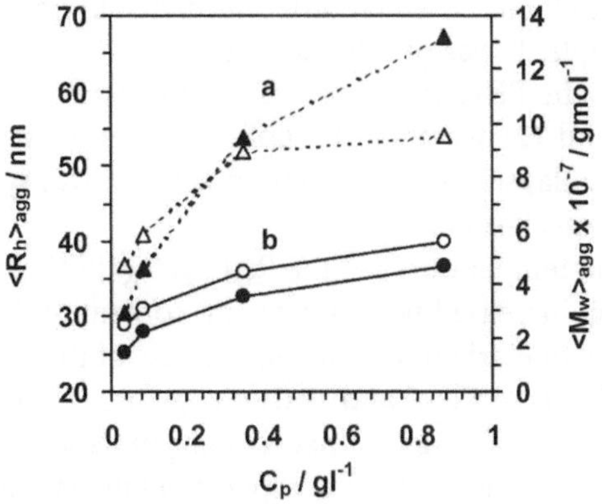

Fig. 7 Concentration dependence of $\langle R_h \rangle$ (*open symbols*) and weight-average molar mass ($\langle M_w \rangle$) (*filled symbols*) of the aggregates at 45 °C: *a* PNIPAM-*g*-PEO-6 and *b* PNIPAM-*g*-PEO-7 [170]

g-PEO copolymers may be taken as examples, have been termed "protein-like copolymers", a concept introduced by Khokhlov et al. [257–260]. The grafting of PEO chains to the copolymer PNIPAM-*co*-GMA was carried out under different conditions to test this hypothesis [165–168, 170]. In aqueous solution, depending on the reaction temperature in the vicinity of the cloud point, the attachment sites for PEO are expected to be located either on the surface of shrunken PNIPAM-*co*-GMA chains ($T \sim$ LCST, PNIPAM-*g*-PEO-7), or randomly throughout the polymer coils ($T \ll$LCST, PNIPAM-*g*-PEO-6), as shown pictorially in Fig. 8. Conditions were selected for which the degree of grafting remained low, compared with the number of reactive groups in

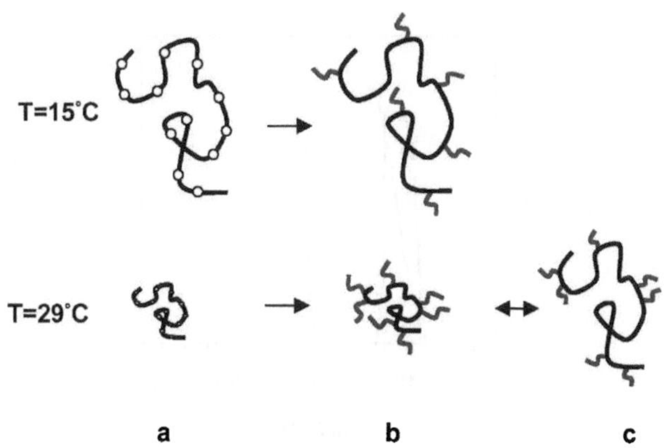

Fig. 8 Model of the grafting of functional copolymers with PEO for reactions carried out at two different temperatures in water: *a* before grafting, *b* after grafting, *c* at room temperature [170]

PNIPAM-*co*-GMA (Table 1). Thus, different distributions of substituents are possible, in principle. It was of interest to see (1) whether the distribution of the PEO grafts on the PNIPAM main chain influences the thermal properties of the polymer and (2) whether the polymer grafted at elevated temperature adopted the collapsed conformation in which it was synthesised when its aqueous solution was heated.

The cloud-point temperatures of the PNIPAM-*g*-PEO samples, determined by light scattering, increased with increasing content of PEO in the polymer, the effect being particularly noticeable in the case of the samples of low grafting density, PNIPAM-*g*-PEO-6 and PNIPAM-*g*-PEO-7. Aqueous solutions of the two copolymers ($c = 1.0\,\mathrm{g\,L^{-1}}$) had cloud points at 32.5 and 33.5 °C, respectively. This difference of 1.0 °C is experimentally significant. The authors took this observation as an indication that the copolymer grafted at elevated temperature is more effectively stabilised by the PEO grafts in the vicinity of the cloud point, compared with the copolymer grafted at lower temperature. The aqueous solutions of these two copolymers were heated fast to 45 °C and the size of the resulting particles was determined. The copolymer synthesised at the higher temperature, PNIPAM-*g*-PEO-7, formed smaller aggregates than PNIPAM-*g*-PEO-6, and the particle size distribution was narrower (Fig. 9). These observations support the hypothesis that the distribution of PEO chains along the PNIPAM chain is different in the two samples and point to the fact that the copolymers tend to remember the conformation in which they were synthesised.

Shrinking of the polymer chains as a function of temperature was observed by capillary viscometry. The reduced viscosity, η_{sp}/c, of polymer solutions approaches zero when the polymers are in their fully collapsed compact state and flow freely through the capillary. Above the cloud point, η_{sp}/c increases, which is indicative of aggregate formation. At low temperatures the

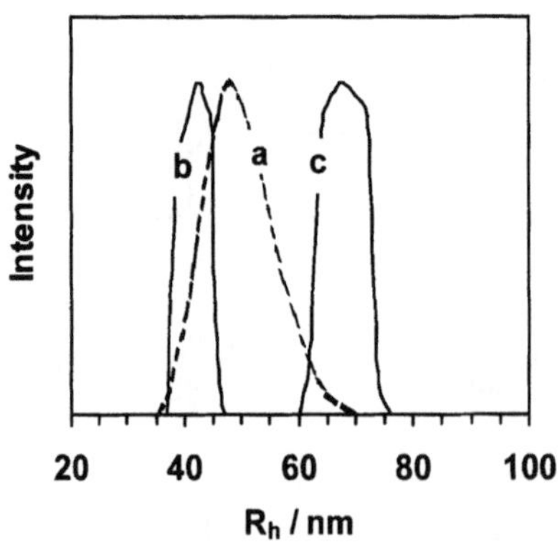

Fig. 9 Size distribution of the aggregates at 45 °C. Polymer concentration $1.0\,\mathrm{g\,L^{-1}}$: *a* PNIPAM-*g*-PEO-6, *b* PNIPAM-*g*-PEO-7, *c* PNIPAM-*g*-PEO-10 grafted in an organic solvent is used as a reference. (Adapted from Refs. [166, 170])

reduced viscosity of the PNIPAM-*g*-PEO-6 is much higher than that of the PNIPAM-*g*-PEO-7. This suggests that the latter polymer adopts a more compact conformation than the copolymer prepared in cold water, pointing to a difference in the distribution of EO grafts on the PNIPAM backbone, as also detected by light scattering measurements (see before). With increasing temperature it is expected that the PEO chains are more localised on the coil surface in PNIPAM-*g*-PEO-7 than in PNIPAM-*g*-PEO-6.

The solubilising effect of PEO chains was also observed by microcalorimetry. The transition temperatures were 33.5 and 34.3 °C for PNIPAM-*g*-PEO-6 and PNIPAM-*g*-PEO-7, respectively. The enthalpy associated with the transition was higher in the case of PNIPAM-*g*-PEO-6 (2.5 kJ mol^{-1} of repeating NIPAM unit) compared with that of aqueous PNIPAM-*g*-PEO-7 (1.3 kJ mol^{-1}). For both copolymers, it was significantly lower than the enthalpy of the phase transition of linear PNIPAM in water (approximately 7 kJ mol^{-1}).

In conclusion, experiments on grafting of PNIPAM with PEO showed that the spatial distribution of PEO grafts along a PNIPAM chain depends on the temperature (below or above the phase-transition temperature) of the aqueous mixture in which the grafting took place if the graft copolymers were synthesised in the vicinity of the phase-separation temperature or far below it. This results in differences in the stability and size of the aggregates formed at elevated temperatures. Below room temperature, the functional PNIPAM derivative is in a coil state and grafting takes place in a random manner. At elevated temperatures, grafting proceeds mainly on the globular surface and results in segmental/nonrandom distribution of the PEO grafts.

2.2.2
PNIPAM-*b*-PEO Copolymers

In this section, we review the properties of a series of PNIPAM-*b*-PEO copolymers with PEO blocks of varying length, with respect to the PNIPAM block. Key features of their solutions will be compared with those of PNIPAM-*g*-PEO solutions. PNIPAM-*b*-PEO copolymers were prepared by free-radical polymerisation of NIPAM initiated by macroazoinitiators having PEO chains linked symmetrically at each end of a 2,2′-azobis(isobutyronitrile) derivative [169, 170]. The polydispersities of PEOs were low, enabling calculations of the number-average molar mass for each PNIPAM block from analysis of their ^{1}H-NMR spectra (Table 2).

Two macroinitiators carrying PEOs of different molar masses ($M_w = 550$ or 1900 g mol^{-1}) were used. By varying the initiator and the initiator-to-monomer ratio, a series of block copolymers was obtained (Table 2). The copolymers (NE) with the longer (1900 g mol^{-1}) PEO block are designated with capital letters A–C, and those with the shorter block (550 g mol^{-1}) with numbers 1–3. Block copolymers having the same PEO block form what is called a copolymer set.

Table 2 Characteristics of PNIPAM-*b*-PEOs [169, 170]

Sample	*n*(NIPA) (mmol)	MAI_{1900} (mol%)[a]	MAI_{550} (mol%)[a]	Solvent (ml)	M_n(PNIPAM) ($g\,mol^{-1}$)[b]	NIPAM/EO[c]
NE-A	17.7	0.01		20	734 000	151
NE-B	17.7	0.1		20	334 000	69
NE-C	8.8	1		10	61 000	13
NE-1	17.7		0.01	20	345 000	244
NE-2	17.7		0.1	20	151 000	107
NE-3	8.8		1	10	39 000	27

[a] Concentrations of macroazoinitiators (*MAIs*) are shown as molar percentages of *N*-isopropylacrylamide (*NIPAM*)
[b] Determined by ^{1}H-NMR
[c] Molar ratio of NIPAM and ethylene oxide (*EO*) repeating units in the blocks

The cloud points of the aqueous polymer solutions with $c = 3 \times 10^{-7}$ M were measured by turbidimetry. Among the copolymer sets, the cloud points of the solutions of block copolymers having the longer PEO block increased from 32 to 34 °C with decreasing molar ratio of NIPAM to EO. No difference between the copolymers having the short PEO block was observed; their cloud point was approximately 33 °C regardless of the NIPAM-to-EO ratio. Thus, the length of the PEO block has a minor effect on the cloud point.

All block copolymers formed aggregates of narrow size distributions when their solutions were brought to 45 °C. The average hydrodynamic radii of the aggregates, $\langle R_h \rangle_{agg}$, increased logarithmically with concentration, whereas their weight-average molar masses, $\langle M_w \rangle_{agg}$, increased linearly with increasing concentration (Fig. 10), unlike PNIPAM-*g*-PEO graft copolymers, for which both $\langle M_w \rangle_{agg}$ and $\langle R_h \rangle_{agg}$ increased logarithmically when c varied (Fig. 7). Within a copolymer set, i.e. copolymers either with a short or a long PEO block, the $\langle R_h \rangle_{agg}$ and $\langle M_w \rangle_{agg}$ values increased with increasing length of the PNIPAM block (and NIPAM-to-EO ration). A comparison of the two copolymer sets, showed that the $\langle R_h \rangle_{agg}$ and $\langle M_w \rangle_{agg}$ values depend more on the ratio of NIPAM to EO than on the length of the PNIPAM block. The only exception was the copolymer with the longest PNIPAM block, which formed the largest aggregates at all concentrations. In dilute solutions ($c < 1 \times 10^{-7}$ M), the $\langle M_w \rangle_{agg}$ values were nearly identical for all block copolymer solutions (Fig. 10b).

The average number of polymer chains per aggregate, $\langle N \rangle_{agg}$, in solutions kept at 45 °C also increased with c. It increased with increasing length of the PNIPAM blocks for copolymers with the longer PEO block, but decreased with increasing PNIPAM block length for the copolymers with the shorter PEO block. The average densities of the aggregates, $\langle \rho \rangle_{agg}$, at 45 °C increased

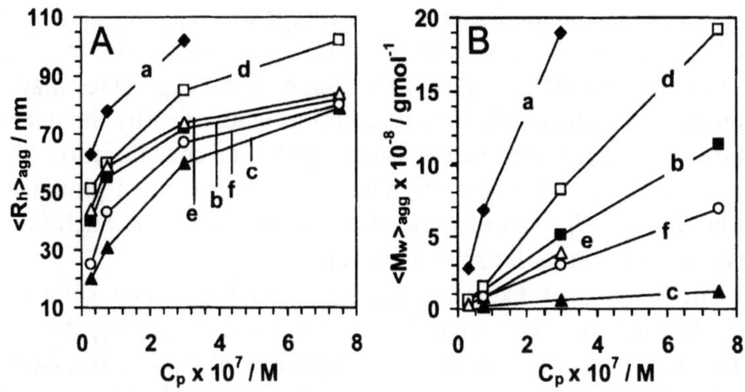

Fig. 10a Average hydrodynamic radii ($\langle R_h \rangle_{agg}$) and **b** weight-average molar mass ($\langle M_w \rangle_{agg}$) of the aggregates in aqueous block copolymer solutions at different polymer concentrations (*c* given in moles of PNIPAM blocks) at 45 °C: *a* NE-A, *b* NE-B, *c* NE-C, *d* NE-1, *e* NE-2, and *f* NE-3. (Reprinted with permission from Ref. [169] copyright 2002 American Chemical Society)

towards a value close to 1 with increasing *c* (Fig. 11). Within a copolymer set, the density of the aggregate decreased with decreasing PNIPAM length and NIPAM-to-EO ratio. The aggregate density of the copolymer having the lowest NIPAM-to-EO ratio remained low over the entire concentration range however, an indication that the high number of EO units, compared with the number of NIPAM units, prevents the effective shrinking and collapse of the PNIPAM segment. Similar behaviour was observed with PNIPAM-*g*-PEO graft copolymers, as described in Sect. 2.2.1.

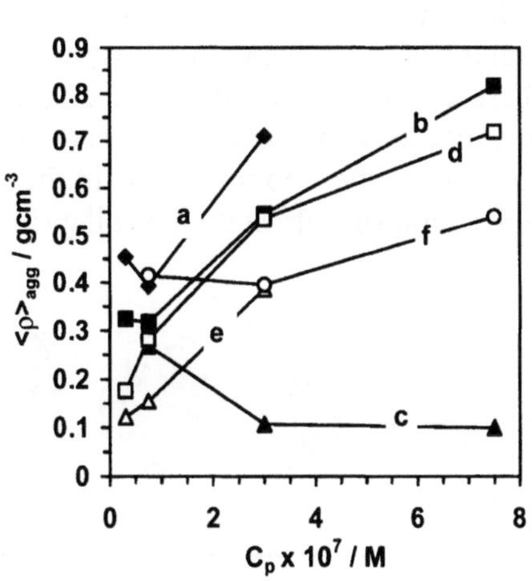

Fig. 11 Average densities of the aggregates ($\langle \rho \rangle_{agg}$) at 45 °C: *a* NE-A, *b* NE-B, *c* NE-C, *d* NE-1, *e* NE-2, and *f* NE-3. (Reprinted with permission from Ref. [169] copyright 2002 American Chemical Society)

The values of the shape factor of the aggregates, $\langle R_g \rangle / \langle R_h \rangle$, changed from around 1.5 (coil-like structure) to around 0.4 with decreasing c or NIPAM-to-EO ratio, a trend indicating that the shape of the aggregates may change from spherical to ellipsoidal. This assumption is justified by the narrow size distribution of the aggregates. For each copolymer concentration, however, the block copolymer of lowest NIPAM-to-EO ratio formed globular structures, whose $\langle R_g \rangle / \langle R_h \rangle$ values were even smaller than 0.78. Such low values are typical of dense core–loose shell structures.

In a further set of experiments, aggregated polymer solutions ($c = 7.5 \times 10^{-7}$ M) kept at 45 °C were diluted with water, keeping the temperature constant. The diluted samples were equilibrated for 1 h, and $\langle R_h \rangle_{agg}$ was measured again. The size of the aggregates formed by the block copolymers remained nearly unaffected by dilution. Thus, the aggregates, once formed, do not disintegrate, implying that the polymer chains are kinetically trapped in the aggregate core, as was also observed in similar experiments carried out with solutions of PNIPAM-*g*-PEO (Sect. 2.2.1).

Block copolymers having closely similar PNIPAM blocks but different PEO block lengths, such as NE-B and NE-1 (Table 2), exhibited distinct aggregation behaviour upon heating. At each concentration, the copolymer with the higher NIPAM-to-EO ratio, NE-1, formed aggregates whose $\langle R_h \rangle_{agg}$, $\langle M_w \rangle_{agg}$, and $\langle N \rangle_{agg}$ were larger than those of NE-B. The average density of the aggregates, however, was almost the same for the two copolymers. Therefore, both size and molar mass of the aggregate are strongly dependent on the length of the PEO block. This is a consequence of the improved steric stabilisation of the aggregates by the longer PEO chains.

The results presented so far suggest that the structure of the aggregates formed at elevated temperature depends both on polymer concentration and on the ratio of NIPAM to EO (Fig. 12). In solutions of low polymer concentration or for low values of the NIPAM-to-EO ratio, the block copolymers form aggregates where the inner part of the core is denser than the average aggregate density, reflecting the extent of the spatial segregation between PNIPAM and PEO. Increasing the amount of PEO in the block copolymer not only has an enhanced solubilising effect on the collapsing PNIPAM block but

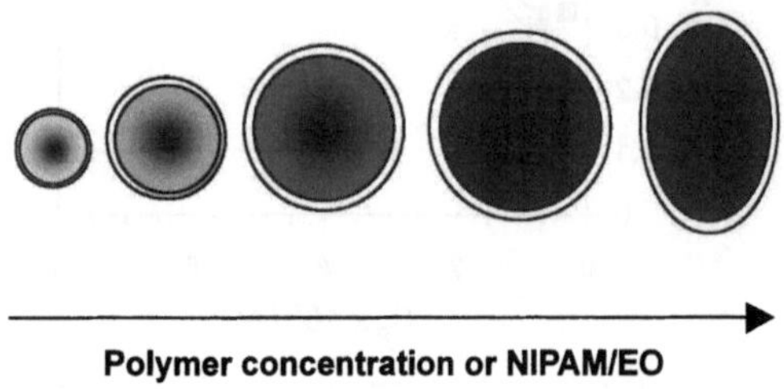

Fig. 12 The evolution of the aggregate with polymer concentration or the molar ratio of NIPAM to ethylene oxide (*EO*) above the LCST [170]

also improves the steric stabilisation of the resulting aggregate. As c or the NIPAM-to-EO ratio increases, the density of the aggregate increases and the mass distribution becomes more homogeneous. At high c values, the block copolymers form aggregates whose shape might change from spherical to ellipsoidal.

It is interesting to describe briefly here the results of two recent studies on the micellisation of NIPAM–EO diblock copolymer of composition other than that of the copolymers discussed before. In one study, using a sample of much lower molar mass $PNIPAM_{44}$-*b*-PEO_{110} (the indices refer to the number of NIPAM and EO units), it was shown that heat-induced micellisation of solutions of high copolymer concentration led to the formation of narrowly distributed, small, dense micelles, whereas large, loose micelles or micellar clusters formed in solutions of low block copolymer concentration [171]. The second recent report relates a thorough comparison of the phase behaviour in water of three samples, $PNIPAM_{228}$-*b*-PEO_{114}, $PNIPAM_{143}$-*b*-PEO_{114}, and $PNIPAM_{38}$-*b*-PEO_{114}, based on a battery of DLS, and small-angle neutron scattering [172]. Phase diagrams within $T = 5–45\,°C$ and $c = 1–70\,g\,L^{-1}$ limits were constructed and were shown to consist of five regions: transparent sol, opaque sol, transparent gel, opaque gel, and syneresis. Microphase separation upon heating already starts at 17–18 °C, when, the authors believe, water starts to act as a selective solvent for PEO blocks and swells the PEO chains at the expense of PNIPAM blocks.

In conclusion, the thermal properties and nanoparticle formation of both block and graft copolymers are governed by several factors. During the collapse, the PNIPAM chains form the core of spherically shaped aggregates surrounded by a hydrophilic PEO shell. As dilute aqueous solutions of block and graft copolymers are heated, the solubilising effect of the hydrophilic segments competes with the hydrophobic interactions amongst shrinking PNIPAM segments. Increasing the number of EO units with respect to NIPAM units results in an elevation of the phase-separation temperature and a slowing down of the collapse of the copolymers. The block copolymers form aggregates of narrow size distribution, whose average size decreases with decreasing polymer concentration. This scenario is also observed in solutions of graft copolymers having a low number of side chains. However, the size distribution of the aggregates of graft copolymers broadens in solutions of copolymers of high degrees of grafting. The size of the aggregates decreases with increasing content of PEO in the case of the block copolymers PNIPAM-*b*-PEO, whereas the average aggregate size of the graft copolymers remains constant, for moderate levels of PEO grafting.

The surface of the PNIPAM-*g*-PEO and PNIPAM-*b*-PEO aggregates is expected to be covered by hydrophilic PEO chains, which impart colloidal stability to the particle. However, some PEO is buried inside the aggregate core. Therefore, increasing mixing of the phases in the core limits core compression. This is especially true in the case of PNIPAM-*g*-PEO. Whereas the block copolymers PNIPAM-*b*-PEO can form very dense aggregates, the core densities of graft copolymers are rather low. The number of polymers in the aggregates formed by the graft copolymers is much lower than that in the aggregates formed by the block copolymers. The difference in the aggregates results from differences in the topology of the block and graft copolymers and in the content of EO repeating units in the two sets of copolymers. In the case of block copolymers, the density of the cores decreases upon dilution of the solution and the mass distribution of the core changes during dilution; the inner part of the core gets more compressed than the rest of the core. In the dilution, the distance between polymers increases, resulting in the domination of the intrapolymeric attractive forces over interchain attraction. This leads to an enhanced surface stabilisation by the hydrophilic parts on the shrunken PNIPAM coil.

In summary, the phase-separation temperature of PNIPAM-PEO block and graft copolymers significantly depends on the amount of PEO and on the polymer concentration. Above the cloud point, the size, size distribution, and structure of the aggregates sterically stabilised by a PEO shell can be adjusted by factors such as the amount of PEO, the polymer concentration, and the heating rate. In the case of graft copolymers, the distribution of PEO grafts along the PNIPAM chain affects the collapse of PNIPAM. Two copolymers grafted in water either far below or close to the cloud point of PNIPAM turned out to be able to partly remember the original conformation of the parent copolymers to which PEO chains were grafted.

2.2.3 Other Copolymers of NIPAM

Several other strategies have been adopted in order to assess the possibility of controlling the architecture of NIPAM copolymers by selecting the appropriate temperature of NIPAM polymerisation in water. An example is the preparation of NIPAM-VP copolymers (5 or 10 mol % of VP) [183]. When the polymerisation was performed in water at a temperature above the phase-separation temperature of PNIPAM, segmented copolymers were obtained, whereas polymerisation conducted in water kept below the cloud point of PNIPAM led to a random distribution of VP units along the chain. The copolymers of segmented VP distribution aggregate more readily upon heating than random copolymers and tend to form stable aggregates larger than those formed by random NIPAM-VP copolymers at elevated temperatures. NIPAM-VP copolymers form aggregates of core–shell structure with VP segments turned towards aqueous surroundings and often called protein-like (Sect. 5.4) [257–260].

The temperature dependence of solutions of an NIPAM–styrene copolymer, PNIPAM-*seg*-St, of $M_w = 13.3 \times 10^6$ g mol^{-1} in which hydrophobic St segments were evenly spaced along the chain was investigated under high dilution conditions, in order to monitor the collapse of this carefully decorated chain [261]. Upon heating, this PNIPAM-*seg*-St chain undergoes two transitions, first from a random coil to an ordered coil in which the hydrophobic stickers tend to gather towards the centre in a more correlated fashion, and then to a collapsed core–shell globule. The value of $R_g/R_h \sim 0.6$ for the globule was significantly lower than the theoretical value, 0.774, corresponding to a uniform hard sphere. This suggests that the single-chain PNIPAM-*seg*-St globule consists of a dense core, presumably made of hydrophobic stickers. No data on the stability against aggregation of these particles were reported in this publication.

A number of diblock copolymers of NIPAM and hydrophobic comonomers have been prepared by various groups and assessed in terms of micellar structure, thermosensitivity, and applications. For example, PS-*b*-PNIPAM was shown to form either micelles consisting of a PS core and a PNIPAM corona, or vesicles. The assemblies were colloidally stable at elevated temperature [262–266].

The solution properties of PNIPAM-*b*-PS and block copolymer of PNIPAM and poly(*tert*-butyl methacrylate) (PNIPAM-*b*-P*t*BMA) have been investigated in water with particular emphasis on the temperature response of these copolymers [185]. Both types of copolymers form colloidally stable micelles in water at room temperature and at elevated temperature, with no sign of aggregation upon heating at 50 °C for several days. The apparent radii of the polymeric micelles at 20 °C were $R_h = 38$ nm and 61 nm, for PNIPAM-*b*-PS and PNIPAM-*b*-P*t*BMA, respectively. The micelles shrunk slightly at 40 °C, the ($R_h = 33$ and 55 nm, respectively). The apparent molar masses of the micelles, extracted from static light scattering measurements, were 10×10^6 g mol^{-1} for PNIPAM-*b*-PS and 164×10^6 g mol^{-1} for PNIPAM-*b*-P*t*BMA at room temperature. They increased to 15×10^6 and 184×10^6 g mol^{-1}, respectively, at 40 °C.

The temperature dependence properties of PNIPAM-*b*-PS and PNIPAM-*b*-P*t*BMA in aqueous solutions heated above the phase-separation temperature of the PNIPAM segment resemble those of PNIPAM itself [141–147]. The PS and P*t*BMA blocks form a dense micellar core, forcing the PNIPAM segments to organise themselves in the shell. This PNIPAM shell totally screens the highly hydrophobic PS and P*t*BMA and controls the solution properties of the micelles below and above the LCST. Thus, it is reasonable to assume that the stabilisation of the micelles above the LCST has the same origin as that of the PNIPAM mesoglobules formed at elevated temperature [141–147]. When two core–shell particles collide above the LCST of PNIPAM, the time of collision is much shorter than the reptation time of the copolymer chains inside each micelle. Indeed the reptation of the copolymer chains is extremely slow, in part as a consequence of the collapsed state of the PNIPAM blocks, but mostly owing to the strong interaction between the PS or the P*t*BMA blocks in the micellar core. Two interacting particles behave like tiny hard spheres and their collision is elastic. In fact, the highly hydrophobic PS and P*t*BMA blocks can be viewed as stabilising agents, i.e. as anchors that prevent copolymer chains from reptating and creating intermicelle entanglements. Supporting evidence is the observation that hydrophobically modified PNIPAM copolymers form smaller mesoglobules than the corresponding homopolymer of similar length and the aggregation number dramatically decreases as the hydrophobic content increases, which might be against our expectations, but can be explained in terms of the viscoelastic effect (Sect. 5.3) [190, 267–270].

The glassy state of the polymers in the mesoglobular phase may be an additional factor affecting the mechanism of micellar stability against precipitation. Thus, the glass-transition temperatures are 106 and 107 °C for PS blocks and 118 °C P*t*BMA [185]. These temperatures are much higher than the range of solution temperatures studied, confirming that reptation of the block copolymers out of the core is restricted. Thermosensitive blocks and homopolymers, as well, form globules with a high-density inner core at elevated temperatures, which may be diagnostic of partial vitrification (Sect. 5.2).

3 Poly(*N*-vinylamides)

3.1 Poly(*N*-vinylcaprolactam)

PVCL is one of several nonionic water-soluble polymers that undergo heat-induced phase separation in water (Fig. 13). It has a repeating unit consisting of a cyclic amide, where the amide nitrogen is connected directly to the hydrophobic polymer backbone.

The first report on the temperature-dependent solubility of PVCL was published in 1968 by Solomon et al. [271]. Up to the 1990s, most studies on PVCL originated from groups in the former Soviet Union. These early studies are reviewed in Kirsh's book [151] on poly(*N*-vinylamides). Recently, PVCL and its solution properties have been scrutinised anew, in view of the noted bio-

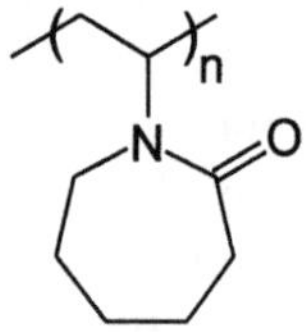

Fig. 13 Poly(*N*-vinylcaprolactam) (*PVCL*)

compatibility of this polymer. Unlike poly(*N*-alkylacrylamides), PVCL does not produce low molecular weight amines upon hydrolysis, since the amide group is part of a caprolactam. In a recent cytotoxicity study of various thermally responsive polymers, Vihola et al. [272] observed that PVCL was well tolerated below and above the phase-transition temperature. This feature, together with its high complexing ability and good film-forming properties, enables the use of PVCL in many industrial applications, in particular in the biomedical field [152, 273].

The phase transition of aqueous PVCL solutions has been investigated over the entire water–polymer composition range. The cloud point of aqueous PVCL solutions lies above 30 °C. It tends to increase with decreasing molar mass [151]. In accordance with the phenomenological classification of Berghmans and Van Mele, PVCL is a type I thermosensitive polymer (Fig. 14) [157, 158]. Polymers of this type follow the classic Flory–Huggins miscibility behaviour, meaning that the LCST shifts towards lower polymer concentrations and temperature upon increasing polymer molar mass.

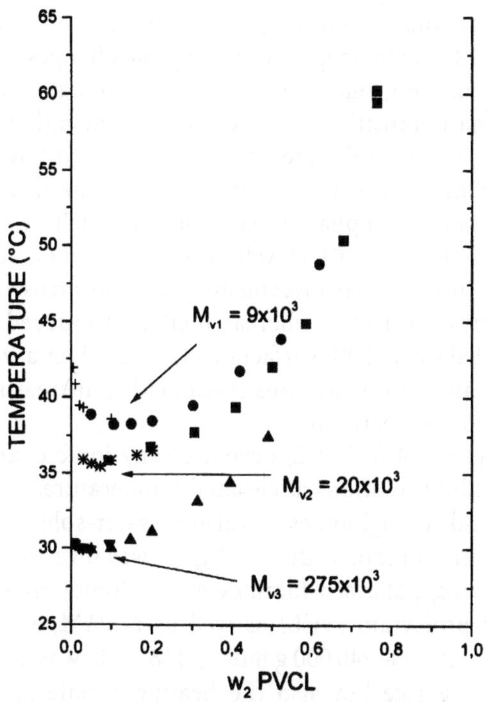

Fig. 14 Type I demixing of PVCL in water for three samples of different chain length (type I). $M_v = 9000\,\mathrm{g\,mol^{-1}}$: T_{cp} (+), T_{dem}, (•); $M_v = 20\,000\,\mathrm{g\,mol^{-1}}$: T_{cp} (∗), T_{dem} (■); $M_v = 275\,000\,\mathrm{g\,mol^{-1}}$: T_{cp} (▼), T_{dem} (▲). (Reprinted with permission from Ref. [157] copyright 2000 Elsevier)

There is no report on direct experimental observations of the coil–globule transition of PVCL single chains. Unlike PNIPAM, PVCL chains exhibit a strong tendency to form aggregates, even in conditions of high dilution, hinting at profound differences in the hydration state of PNIPAM and PVCL. Nonetheless, numerous attempts have been reported. Thus, observation of the coil–globule transition of PVCL ($M_w \sim 10 \times 10^6$ g mol^{-1}) in D_2O by small-angle neutron scattering has recently been presented [274, 275]. The study was performed at the threshold of the polymer overlap concentration c^*. Lau and Wu [276] monitored the temperature dependence of the hydrodynamic radius, R_h, of a high molecular weight PVCL ($M_w \sim 10 \times 10^6$ g mol^{-1}) in water by light scattering techniques in the very dilute regime ($c < 1$ g L^{-1}). Experiments were performed in the vicinity of the phase-transition boundary. They observed that R_h decreases as the solution temperature nears the phase-separation temperature, indicating a contraction of the polymer chain. The aggregation of the chains, signalled by an increase in the scattering intensity, was detected at a temperature in the vicinity of 31 °C. Makhaeva et al. [277] noticed that in solutions of PVCL containing ionic surfactants, an increase in temperature results in a sharp drop of the hydrodynamic diameter of the PVCL coil for a given critical temperature. Further insight into the water–PVCL interactions in solutions below and above the LCST is given in a recent study, by IR spectroscopy monitoring the changes in hydrogen-bond formation of the amide groups and the hydration states of the alkyl moieties during the phase transition [278]. The hydration and phase behaviour of PVCL and poly(*N*-vinylpyrrolidone), PVP, in water and D_2O were compared using DSC, turbidimetry and Fourier transform IR spectroscopy. PVP solutions in 1.5 M KF undergo a phase separation around 30 °C. The effect of the addition of KF, KCl, KBr, and KI as well as of methanol on aqueous solutions of PVCL and PVP in water was investigated by IR spectroscopy. Salts did not alter the IR spectrum of PVCL, whereas methanol induced a change in the amide I band. Added salts [278], surfactants [277, 279], or a cosolvent [278] all affect the phase transition of aqueous PVCL, either increasing or decreasing the phase-separation temperature.

Like PNIPAM [141–145], PVCL, devoid of any ionic charge, forms stable mesoglobules in dilute solutions at elevated temperatures [146, 147]. Formation of globules and mesoglobules of various water-soluble thermosensitive polymers at 60 °C in solutions with $c < 0.1$ g L^{-1} was investigated by Anufrieva et al. [280, 281] using polarised luminescence. Homopolymers and copolymers of *N*-vinylcaprolactam (VCL) as well as NIPAM, and NIPMAM were synthesised ($M_w = 20\,000–240\,000$ g mol^{-1}) [281]. It was demonstrated that not only the heating rate but also the heating profile affect the size and composition of mesoglobules [281]. Thus, an aqueous PVCL solution was heated to 60 °C and then quickly cooled to 8 °C, kept at this temperature for 1 h, and subsequently reheated to 60 °C. The authors performed this experiment to test the hypothesis that in solutions kept at 8 °C intermolecular contacts diminish, whereas intramolecular contacts within the mesoglobular core remain undisturbed. This procedure is in fact similar to the fast heating/quenching applied in Ref. [147]. It was observed that these conditions lead to the formation of mesoglobules consisting of four to five macromolecules in the case of PVCL of $M_w < 50\,000\,\mathrm{g\,mol^{-1}}$. Solutions of PVCL slowly heated from room temperature to 60 °C formed larger aggregates.

The structure of PVCL mesoglobules at 50 °C can be frozen by physically cross-linking the collapsed PVCL chains with multifunctional phenols able to form hydrogen bonds with the VCL amide groups [146]. This leads to nanosized hydrogel particles that are stable even at room temperature [180]. This cross-linking of thermally induced PVCL particles provides a method to produce well-defined nanosized hydrogel particles that may meet various biotechnology needs, such as enzyme stabilisation or controlled drug release, in view of the biocompatibility of PVCL [272].

3.2
Linear and Cross-Linked Derivatives of PVCL

In 1958, Freedman et al. [282] reported the first synthesis of vinyl monomers, which also functioned as emulsifying agents. Since then, a great variety of reactive surfactants, ionic [283–285] and nonionic [286, 287], have been prepared and used in heterogeneous polymerisation processes. Asua et al. [288] have reviewed the general features of emulsion and dispersion polymerisation mechanisms. Nonionic reactive surfactants having a molar mass of 10^3–$10^4\,\mathrm{g\,mol^{-1}}$ are usually referred to as "amphiphilic macromonomers". Typically, such macromonomers consist of a PEO segment, the hydrophilic part, linked to an alkyl chain, the hydrophobic part. Other types of macromonomers have been synthesised as well [289]. These macromonomers are mainly used in emulsion and dispersion polymerisations to replace conventional surfactants.

Novel thermally responsive polymeric nanoparticles that show high colloidal stability were prepared by Laukkanen et al. via emulsion polymerisation of VCL in the presence of the amphiphilic macromonomer ω-methoxy poly(ethylene oxide)$_{42}$ undecyl α-methacrylate ($MAC_{11}EO_{42}$) (Fig. 15) [177, 178, 182]. The macromonomer itself proved to be highly sur-

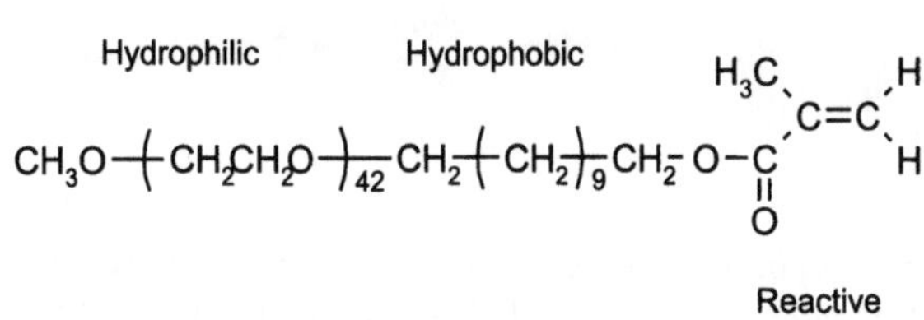

Fig. 15 Amphiphilic PEO-alkyl macromonomer $MAC_{11}EO_{42}$

face active and prone to form micellar structures in an aqueous environment. The resulting microgels proved to be colloidally very stable, even in concentrated electrolyte solutions, which indicates an effective steric stabilisation. They underwent heat-induced reversible volume change from expanded to shrunken states (Fig. 16). A detailed account of the properties of these macrogels is presented in Sect. 3.2.1.

Section 3.2.2 is devoted to a discussion of the synthesis and solution properties of various graft copolymers of VCL and $MAC_{11}EO_{42}$ obtained by carrying out conventional solution polymerisations in organic solvents [179, 180, 182]. By changing the ratio between reactive and nonreactive surfactant it was possible to achieve graft copolymers with different degrees of grafting. Below the LCST, the graft copolymers form intrapolymeric and interpolymeric associates in water depending on polymer concentration and grafting density. The associates are composed of a hydrophobic nonpolar core surrounded by self-assembled $C_{11}EO_{42}$ grafts. Upon heating, the PVCL backbone collapses, triggering a change in the hydration of the chain and release of polymer-bound water molecules. Thermally induced aggregation leads to the formation of colloidally stable mesoglobules (Fig. 16). The graft copolymers form less dense particles than PVCL homopolymers. Each particle is composed of several thousands of collapsed polymer chains and the particles still contain much water at 50 °C. Changing the concentration or the composition of the graft copolymer alters the size of the mesoglobules.

Using this macromonomer technique, a great variety of polymeric architectures can be attained, such as comblike, brushlike, and starlike copolymers [289–291]. Particularly interesting structures are achieved when a hydrophobic monomer is copolymerised with a hydrophilic macromonomer, or vice versa. The resulting graft copolymers show a wide range of solution properties in water, owing to the varying hydrophilic–hydrophobic balance. For example, hydrophobically modified water-soluble macromolecules associate in water, forming hydrophobic microdomains [29]. These assemblies impart both dilute and semidilute solutions of unique properties, which have found numerous industrial applications [293–295].

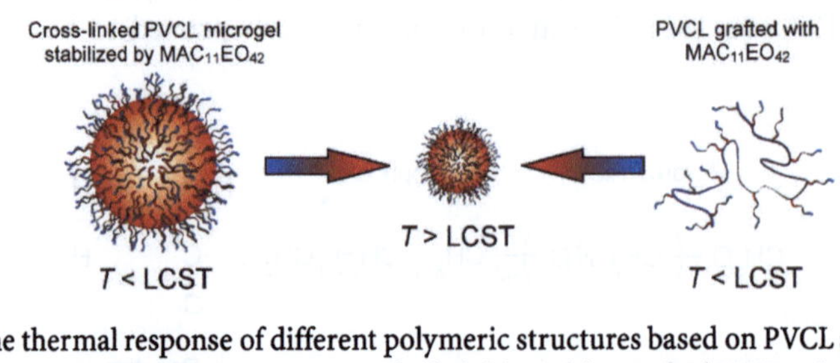

Fig. 16 The thermal response of different polymeric structures based on PVCL and the amphiphilic macromonomer $MAC_{11}EO_{42}$. *Left*: Shrinking of the grafted PVCL microgel. *Right*: Heat-induced aggregation of the graft copolymer and formation of a mesoglobule [181]

3.2.1 Cross-Linked PVCL Microgels Stabilised with Amphiphilic Grafts

Laukkanen et al. [177, 178, 182] used emulsion polymerisation to prepare cross-linked PVCL microgels. The experimental details are outlined in Table 3. Special attention was paid to controlling the surface properties of the microgels. Four different PVCL microgels were prepared by specific selection of the initiators and surfactants. Negatively charged particles were obtained using an ionic initiator, potassium persulphate (KPS) which upon decomposition forms sulphate anion radicals, which covalently bind to the growing polymer. Ionic surfactants, such as SDS, which tend to bind to the polymers, also introduce electric charges, since they cannot be totally removed even by thorough purification of the polymer [296]. Nonionic microgel particles were prepared in emulsions using an electrically neutral initiator and surfactant. Grafted particles were obtained via postmodification of PVCL microgels with a macromonomer.

The size distribution of the PVCL microgel particles synthesised by a batch emulsion polymerisation was monomodal and reasonably narrow [177], regardless of the choice of the emulgator, SDS (E1, E2) or macromonomer (E3, E4). The size distributions remain monomodal upon subsequent grafting. A typical size distribution of a PVCL microgel (E1) at 20 °C is presented in Fig. 17, as well as the effect of grafting, as a second step, on the hydrodynamic size (E1-*g*).

The sizes of charged and neutral particles were measured as a function of increasing temperature (Fig. 18) [178]. The thermal collapse of the PVCL particles turned out to be a more or less continuous process, regardless of particle charge and the presence or absence of amphiphilic grafts. The PEO chains bound to the particle surfaces had only a minor effect on the transition temperature. The sizes of the particles in cold water varied from sample

Table 3 Summary of the reaction conditions in emulsion polymerizations. (Adapted with permission from Ref. [178] copyright 2002 Springer-Verlag)

Sample	VCL (g/L^{-1})	Surfactant (g/L^{-1}) SDS	$MAC_{11}EO_{42}$	Initiator (g/L^{-1}) KPS	VA-086	Temperature (°C)	Reaction time (h)
E1	6.70	0.42	–	0.25	–	70	20
E2	6.70	0.42	–	–	2.69	75	20
E3	6.70	–	2.94	0.25	–	70	20
E4	6.70	–	2.94	–	2.69	75	20

VCL *N*-vinylcaprolactam, *SDS* sodium dodecyl sulphate, $MAC_{11}EO_{42}$ ω-methoxy poly-(ethylene oxide)$_{42}$ undecyl α-methacrylate, *KPS* potassium persulphate, *VA-086* 2,2′-azobis[2-methyl-*N*-(2-hydroxyethyl)propionamide]

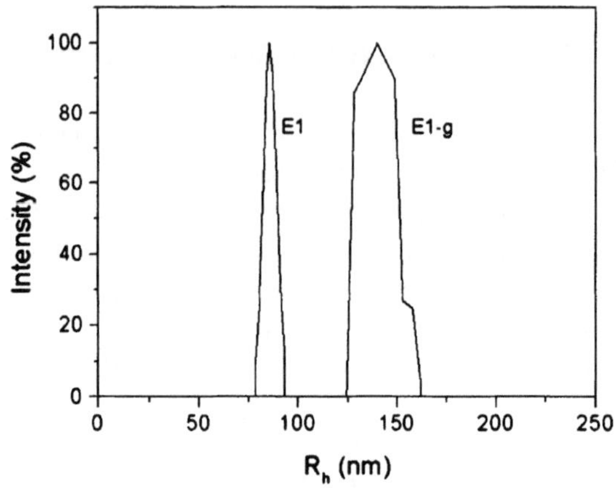

Fig. 17 Size distributions of PVCL particles before (*E1*) and after (*E1-g*) grafting in water at 20 °C. (Adapted from Refs. [177, 181])

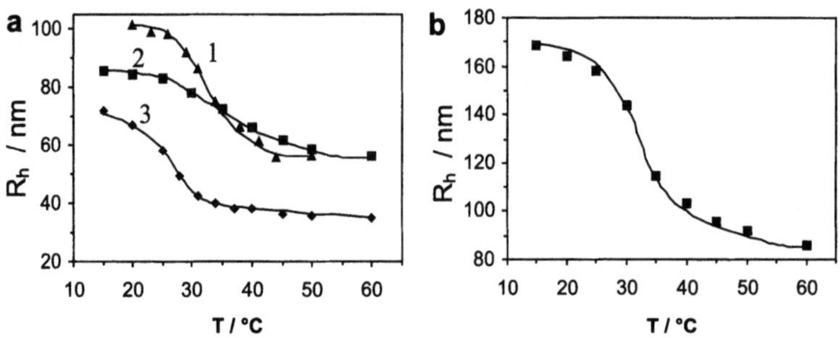

Fig. 18 **a** Temperature dependence of the hydrodynamic radius of charged PVCL microgels E1 (1), E2 (3), and the grafted E3 (2). **b** Temperature dependence of the R_h of the electroneutral grafted PVCL microgel E4. (Reprinted with permission from Ref. [178] copyright 2002 Springer-Verlag)

to sample. The differences in particle sizes may originate from several factors: (1) the effectiveness of the various surfactants in stabilising the growing polymer particles; (2) ionic initiators are known to generate oligomeric radical anions which may act as surfactants; (3) different initiators have different dissociation rates, a property which affects the number of polymer particles produced and, thus, the rate of polymerisation. In all cases, the hydrodynamic radius of the particles decreased to approximately half its original value upon heating a particle suspension above the LCST.

The introduction of amphiphilic macromonomer chains on the surface of PVCL microgels did not affect the thermal properties of the PVCL microgels: grafted particles gradually collapsed upon heating within a temperature range identical to that recorded in the case of suspensions of the precursor particles. Also, the degree of shrinking was approximately the same in all case: the hydrodynamic radii of particles at 20 °C were twice as large as those of particles heated to 50 °C.

The mobility of the amphiphilic graft was followed by NMR diffusion measurements over a broad temperature range. The relaxation times τ_1 of the CH_2 protons of the PEO segments were measured for a micellar solution of $MAC_{11}EO_{42}$ and for suspensions of various $MAC_{11}EO_{42}$-grafted microgels, kept below and above their volume-transition temperature. The relaxation time increased in all cases with increasing temperature and the relaxation profiles were found to fit a single-exponential law. The relaxation behaviour did not change when the suspension temperature neared 35 °C, implying that the dynamics of the PEO segments is not affected by the collapse of the PVCL chains. The grafts were assumed to adopt a brush conformation, similar to their conformation in macromonomer micelles both below and above the cloud point of PVCL, or on the surface of latex particles [244], The diffusion coefficients of the CH_2 protons in the PEO segments increased slightly with increasing temperature (Fig. 19) [177]. This indicates either that macromonomers adsorbed on the particle surface tend to dissolve in the aqueous phase during the collapse transition or that a significant number of the macromonomers exist as micelles in the mixture.

PVCL microgels prepared via covalent binding of PEO exhibit different temperature dependence (Fig. 19). In this case, a considerable increase in the diffusion coefficient takes place above the LCST of PVCL. The sudden increase may be attributed to the shrinking of the particle, which leads to an increase in the rate of its translational diffusion and, consequently, also in the rate of diffusion of the grafts bound to the particle surface. The values of the diffusion coefficients above the LCST should be taken as apparent ones, as the measurements were complicated by the heterogeneity of the collapsed samples.

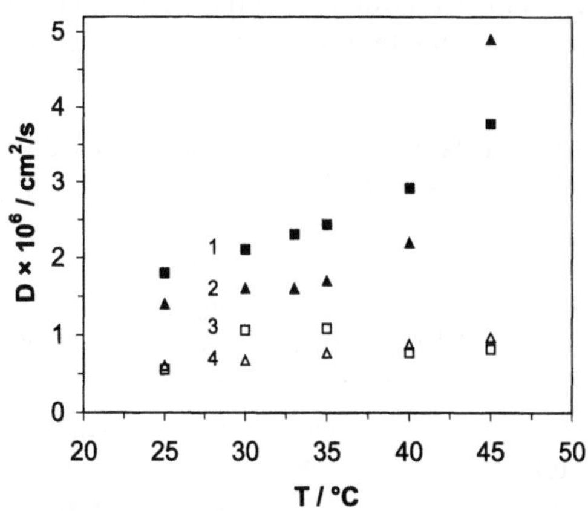

Fig. 19 Diffusion coefficients of the samples: $MAC_{11}EO_{42}$-grafted microgels (1, 2), mixture of PVCL microgel E1 and the macromonomer (3), micellar solution of macromonomers (4). (Adapted from Refs. [177, 181])

The colloidal stability of the two types of microgels arises from different origins: $MAC_{11}EO_{42}$-grafted microgels are expected to be sterically stabilised against coagulation by the grafted PEO chains, whereas unmodified particles are stabilised electrostatically. Owing to the choice of initiator and surfactant in the emulsion polymerisation, different charge densities were expected to appear on the surface of the particles. Electrophoretic mobilities of the PVCL particles were measured at two temperatures, 25 and 45 °C, to determine the surface charge introduced by the ionic initiator, KPS, and the surfactant, SDS.

The zeta potentials of particles prepared by the two routes differed noticeably (Table 4). Particles obtained in the presence of either KPS or SDS have a slightly negative zeta potential at 25 °C (E2, E3). Particles synthesised using nonionic initiator and surfactant (E4) had zero charge. The electrophoretic mobility of the negatively charged particles increased with increasing temperature, as a result of the collapse of the particles and of the decrease in the viscosity of the electrolyte solution. The increase of the electrophoretic mobility owing to the collapse may be caused not only by the reduction of the particle size but also by an increase in the surface fraction of charged groups, when the water content inside the particle decreases. Similar phenomena were observed with other responsive particles [297, 298]. The zeta potentials of samples E1 and E2 differed noticeably at 25 °C but reached almost the same value at 45 °C.

The stability of the responsive microgel particles was studied as a function of added electrolyte concentration [178]. The changes in turbidity of sample E2 (a charged microgel) and sample E4 (an electroneutral $MAC_{11}EO_{42}$-grafted microgel) as a function of barium chloride concentration are presented in Fig. 20. At 21 °C both samples are stable, with no indication of coagulation in the presence of salt, up to a concentration of 0.4 mol L^{-1} in the case of E2, and even higher for E4 (Fig. 20a).

At 45 °C (Fig. 20b), the situation is different. The electrostatically stabilised particles (E2) start to coagulate and form large aggregates in the presence

Table 4 Summary of the capillary electrophoretic study.(Adapted with permission from Ref. [178] copyright 2002 Springer-Verlag)

Sample	$T = 25$ °C			$T = 45$ °C		
	R_h (nm)	μ_e (10^{-8} m^2 V^{-1} s^{-1})	ζ (mV)	R_h (nm)	μ_e (10^{-8} m^2 V^{-1} s^{-1})	ζ (mV)
E1	98	−1.27	−16.1	57	−2.10	−17.4
E2	58	−0.44	−5.6	36	−1.97	−16.3
E3	83	−0.24	−3.0	62	−6.46	−5.4
E4	159	–	–	96	–	–

μ_e electrophoretic mobility, ζ zeta potential

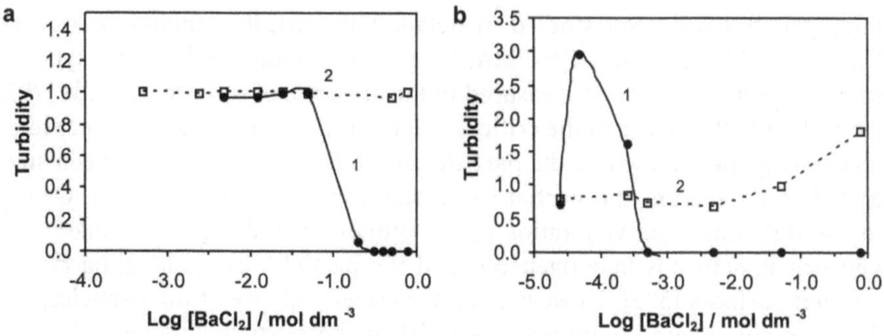

Fig. 20 Effect of $BaCl_2$ on the turbidity of microgel dispersions at **a** 21 °C and **b** 45 °C: E2 (1) and E4 (2). (Reprinted with permission from Ref. [178] copyright 2002 Springer-Verlag)

of very moderate electrolyte concentration. They precipitate upon centrifugation. The first stage of aggregation can be detected by the very abrupt increase in the relative turbidity at low electrolyte concentration (Fig. 20b). The critical coagulation concentration for E2 at 45 °C was in the range $5 \times 10^{-5} - 1 \times 10^{-3}$ mol L^{-1}. The sterically stabilised particles (E4) did not precipitate during centrifugation and the increase in turbidity in the early stage of aggregation was only moderate even for very high electrolyte concentrations, i.e. above 0.05 M.

According to the DLVO theory, the electrostatic repulsion between particles due to the electric double layer decreases with increasing ionic strength of the medium. The effect depends on the charge of the cation; the higher the charge, the higher the reduction in the repulsive potential between the particles. Added electrolytes also strongly affect the solubility of PEO and decrease its ability for steric stabilisation [299, 300]. The effect of electrolytes on the solubility of PEO is also dependent on the charge of the cation; by decreasing the valency of the cation, the solubility of PEO decreases [299]. To understand the mechanism of stabilisation of PEO-grafted particles, one needs to take into account these two opposing effects. Barium chloride strongly affects the electric double layer but has only a minor effect on the solubility of PEO at low temperatures and low electrolyte concentrations [6, 299]. A concentrated electrolyte solution may, however, be a poor solvent for PEO. In such a case, the thickness of the stabilising PEO shell decreases, and bridging flocculation may occur [3]. For example, aqueous 0.39 M $MgSO_4$ solution is a Θ-solvent for PEO above 40 °C [299, 300]. This might be the reason for the slight instability of the $MAC_{11}EO_{42}$-grafted microgels at very high electrolyte concentrations.

There are several reports on the stability of PEO-grafted latex particles synthesised using the macromonomer in highly concentrated electrolyte solutions [301, 302]. However, the stability of thermally responsive PEO-grafted microgels has not been studied in detail. The critical coagulation concentrations of PNIPAM and PNIPMAM microgels stabilised by carboxylic or sulphate groups have been measured in the presence of sodium chloride [297, 298]. Usually the values of the critical coagulation concentration are very high at low temperatures, where the particles are fully swollen. This has been suggested to arise from the partial steric stabilisation of the particles. Owing to the different reactivity ratios of the monomer and the cross-linker, the network structure is heterogeneous and the particles carry dangling chains on their surfaces [303]. However, upon thermal collapse of the particles, the dangling chains are compressed, parallel to the particle surface and, thus, steric stabilisation ceases to be effective. Indeed, it has been shown that loose, porous PNIPAM particles turn into hard spheres upon heating above the collapse temperature [304]. At elevated temperatures, electrosterically stabilised PVCL microgels have properties similar to those of PNIPAM microgels, in particular in their behaviour towards added salts [297].

As a conclusion, grafting the responsive PVCL particles with PEO derivatives increased their colloidal stability in the presence of added electrolytes, the effect being especially pronounced at high temperatures where the PVCL particles were in their shrunken state. In the absence of stabilising grafts, repulsive forces between negatively charged PVCL particles decreased sharply upon addition of an electrolyte, as expected according to the DLVO theory. It may be concluded that the macromonomer technique is the method of choice for the synthesis of thermally responsive PVCL microgels that neither coagulate nor precipitate in solutions of high ionic strength.

3.2.2
Amphiphilic Graft Copolymers of VCL

Various copolymers of VCL and the amphiphilic macromonomer $MAC_{11}EO_{42}$ were synthesised by solution polymerisation in benzene [179, 180, 182]. The macromonomer content varied from 0 to 34 wt % and the molecular masses of the copolymers were on the order of 300 000 g mol^{-1} (Table 5). Several authors have reported that PEO–methacrylate macromonomers can act as chain transfer agents, leading to branching and cross-linking during the polymerisation [305, 306]. The molar mass distributions of the VCL-$MAC_{11}EO_{42}$ were monomodal with no evidence of cross-linking. The long alkyl chain between the reactive methacrylate group and the PEO segment may account for the difference in behaviour between PEO–methacrylates and $MAC_{11}EO_{42}$. PVCL samples with M_w varying from 21 000 to 1 500 000 g mol^{-1} were also synthesised as model polymers (Table 6).

Microcalorimetric endotherms recorded for solutions of PVCL and the PEO-grafted PVCL copolymers are broad, and markedly asymmetric, with a sharp increase of heat capacity on the low temperature side (onset of the transition, T_{dem}) and a more gradual decrease of the heat capacity as the tem-

Table 5 Summary of the reaction conditions for poly(*N*-vinylcaprolactam) (*PVCL*) grafted with $MAC_{11}EO_{42}$ [181]

Sample	VCL ($mol\,L^{-1}$)	$MAC_{11}EO_{42}$ ($mmol\,L^{-1}$)	AIBN ($mmol\,L^{-1}$)	M_w ($g\,mol^{-1}$) [a]	$MAC_{11}EO_{42}$ content (wt %)
PVCL-*g*-6	1.08	3.6	2.56	71 000	6.3
PVCL-*g*-13	1.08	7.1	2.56	310 000	13.0
PVCL-*g*-16	1.08	10.9	2.56	250 000	15.8
PVCL-*g*-18	1.08	14.5	2.56	300 000	18.3
PVCL-*g*-34	1.08	21.3	2.56	260 000	34.0

AIBN 2,2′-azobis(isobutyronitrile)
[a] The molecular weights were measured with static light scattering and should be taken as apparent ones.

perature increases beyond the maximum temperature T_{max} (Fig. 21) [181]. As discussed in Sect. 2.1, T_{dem} is closely related to the onset of aggregation and corresponds well to the cloud-point temperatures of the solutions.

Turning our attention first to the endotherms recorded for PVCL (Fig. 21), we note that the onset temperature of the transition, T_{dem}, and the temperature of the transition, T_{max}, depend on the molar mass of the polymer: T_{dem} decreases with increasing molar mass. The heat of the transition, however, does not depend on the molar mass of the PVCL chain, $\Delta H = 4.4 \pm 0.4\,kJ\,mol^{-1}$ [182]. The introduction of amphiphilic grafts along a PVCL chain shifts T_{dem} and T_{max} to higher temperature, compared with the temperatures

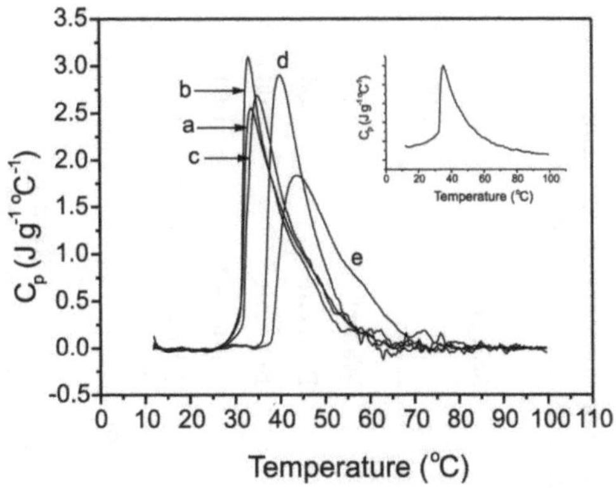

Fig. 21 Microcalorimetric endotherms for aqueous solutions of PVCL samples. PVCL-1500 (*a*), PVCL-1300 (*b*), PVCL-330 (*c*), PVCL-30 (*d*), PVCL-21 (*e*). An endotherm prior to baseline subtraction is shown for PVCL-330 in the *inset*. (Reprinted with permission from Ref. [146] copyright 2004 American Chemical Society)

Table 6 Summary of the reaction conditions and molecular characteristics of the PVCL homopolymers. Adapted with permission from Ref. [146] copyright 2004 American Chemical Society)

Sample	Solvent	VCL ($mol\,L^{-1}$)	Initiator ($mmol\,L^{-1}$)	Temperature (°C)	Reaction time (h)	M_w ($g\,mol^{-1}$)[c]	Polydispersity index[d]
PVCL-1500	Water[a]	1.08	7.28	70	20	1 500 000	2.3
PVCL-1300	Benzene[b]	1.08	7.03	35	68	1 300 000	1.8
PVCL-330	Benzene[b]	1.08	2.56	70	20	330 000	1.6
PVCL-30	Toluene[b]	1.08	2.56	70	20	30 000	1.4
PVCL-21	2-Propanol[b]	1.08	2.56	70	20	21 000	1.7

[a] Initiator VA-086
[b] Initiator AIBN
[c] Measured by static light scattering
[d] Measured by size-exclusion chromatography

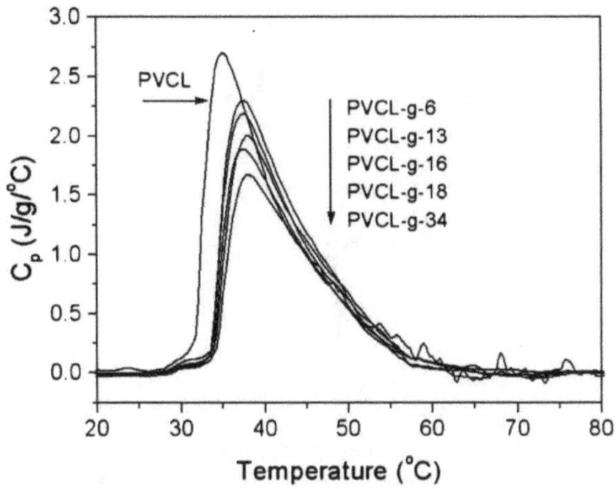

Fig. 22 Microcalorimetric endotherms for an aqueous solution of PVCL-330 and the corresponding graft copolymers. Polymer concentration $1\,g\,L^{-1}$ and heating rate $60\,°C\,h^{-1}$. (Reprinted with permission from Ref. [180] copyright 2005 Elsevier)

for PVCL (Fig. 22). The T_{dem} and T_{max} values are independent of the degree of grafting. The heat of the transition, ΔH, in contrast, depends on the graft content of the copolymers, but the enthalpy change calculated per VCL repeating unit remains constant.

In the inset of Fig. 21, a calorimetric trace recorded for a solution of PVCL is presented prior to baseline subtraction. One notes in this thermogram that the heat capacity after the transition is lower than that before the transition. This was the case for all polymers: ΔC_p was $-70 \pm 20\,J\,mol^{-1}\,°C^{-1}$. The same effect was observed during cooling scans from 100 to 10 °C, though the sign of ΔC_p was opposite. A similar negative change in C_p upon heating was also observed for PNIPAM ($\Delta C_p = -63\,J\,mol^{-1}\,K^{-1}$) [209] as well as for the phase transitions of various pluronic-type block copolymers [307]. It was taken as an indication of diminished interaction between water molecules and polymer chains.

Additional information on the solvation layer around the polymers was obtained by PPC (Sect. 2.1), a technique that allows one to evaluate the changes in the partial volume of the polymer throughout the phase transition, and to obtain information on the temperature-dependant relative hydrophilicity/hydrophobicity of a polymer in solution [210]. Particular interest in the PPC studies was focused on the effect of the amphiphilic grafts on the volumetric properties of the polymers.

The thermal expansion coefficients of PVCL and a copolymer in water, α_{pol}, were determined by PPC as a function of temperature (Fig. 23) [180]. The plots can be divided into four temperature ranges. Below the transition temperature, $10 < T < 30$ °C, α_{pol} for PVCL remains constant, while in the case of PVCL-*g*-34, α_{pol} has a negative slope. In both cases, α_{pol} undergoes a sharp

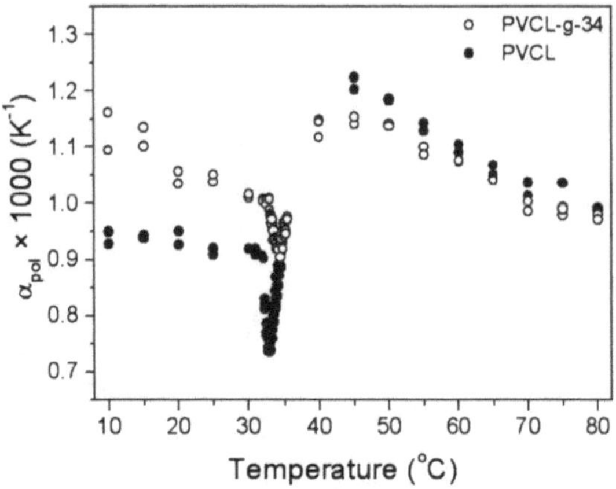

Fig. 23 Temperature dependence of the coefficient of thermal expansion (α_{pol}) of PVCL in H_2O (•) and in PVCL-*g*-34 (○). (Reprinted with permission from Ref. [180] copyright 2005 Elsevier)

decrease, reaches a minimum at 33–34 °C, then increases abruptly with increasing temperature to reach a maximum value at $T \sim 45$ °C, and it gradually decreases as the temperature further increases. The plots recorded for the other $C_{11}EO_{42}$-grafted PVCL samples present features similar to those of the copolymer with highest graft density. Integration of $\alpha_{pol}(T)$ yields $\Delta V/V$, the percentage change in the hydration volume of the partial volume of the polymer as the polymer chain collapses. The volume change is usually taken as the area defined by the peak of the PPC scan and a progress baseline drawn from projections of the baselines in the pretransition and posttransition regions [210]. Application of this method to the PPC scan of PVCL in water yields a value of $\Delta V/V = -0.10\%$ for the sharp negative transition at 32.9 °C (Fig. 23). However, the PPC trace recorded for PVCL solutions may also be seen as consisting of two parts, the sharp signal ($\Delta V/V = -0.10\%$) and a broader signal ($\Delta V/V = +0.50\%$) in the low and high temperature ranges of the transition, respectively. Correspondingly, similar values, although smaller in magnitude, are obtained for the graft copolymers. This data analysis is supported by the asymmetry of the DSC trace (Figs. 21, 22), which may be interpreted as the overlap of two phenomena, first the collapse of the chains, then the aggregation of the collapsed chains, both phenomena characterised by a change in the polymer hydration volume.

The $\alpha_{pol}(T)$ curves for the homopolymer and the graft copolymer differ in one aspect, the slope in the pretransition region ($10 < T < 30$ °C). The α_{pol} value is nearly constant in this temperature domain for PVCL, whereas in the case of the graft copolymer it significantly decreases with increasing temperature. High positive values of the thermal expansion coefficient and strong temperature dependence of α are typical characteristics of molecules that act as structure breakers in water, such as the polar hydrophilic amino acids asparagine and glutamine [210]. In the case of $C_{11}EO_{42}$-grafted PVCL, the high PEO content seems to cause a similar effect: PEO acts as a structure breaker at temperatures between 10 and 30 °C. At elevated temperatures ($45 < T < 80$ °C) the slope of $\alpha_{pol}(T)$ remains negative, indicating that surfaces of the structures generated by the collapsed polymers may still be partially hydrophilic even though the polymer has expulsed the majority of the water molecules.

The magnitude of the volume change for PVCL and its grafted derivative is largest for the homopolymer. The reduced changes in the hydration volume in the cases of graft copolymers may indicate that there is less "ordered water" bound to the grafted polymer chains, possibly owing to the structure-breaking properties of the hydrophilic PEO chains and, hence, a smaller volume of directly interacting water. This trend is opposite to that exhibited by hydrophobically modified PNIPAMs, which undergo larger volume changes as a result of the introduction of structure-making groups [308].

The main features of the PPC curves can be rationalised from an understanding of the phase-separation phenomenon of the PVCL chain. Below the phase-transition temperature, the PVCL chains are solvated to the greatest extent, implying a large number of contacts between the repeating units and water molecules, a fact confirmed by IR spectroscopy studies of cold (5 °C) PVCL solutions [309]. Other weak forces also occur in the water–PVCL system, namely dipole–dipole interactions between the polar amide groups and dispersion forces between the methylene groups of the caprolactam rings in the chair conformation. The hydration of the hydrophobic methylene groups is expected to weaken mildly as the temperature increases [31, 310]. Several events take place at the phase transition. Contacts between water molecules and the hydrophobic groups of the polymer become thermodynamically less favourable than contacts between the hydrophobic groups themselves. As a result, the polymer chains tend to collapse into a conformation in which the methylene groups are mostly shielded by hydrophilic polar groups of the polymer, resulting in a decrease of the solvent accessible surface area and, consequently, of the hydration volume, detected in the PPC experiment as a small negative signal. Owing to topological constraints, all the hydrophobic groups cannot be shielded from surrounding water molecules, and eventually, with further increase of the temperature, collapsed chains tend to aggregate, releasing bound water molecules, a mechanism responsible for the positive volume change detected by PPC.

The association of the graft copolymers in water at room temperature was studied by DLS [180]. The hydrodynamic radius of the various copolymers in dilute aqueous solutions at 20 °C, well below the cloud point, was determined for solutions ranging in concentration from 1 to 10 $g\,L^{-1}$. Similar measurements were also conducted for solutions of an unmodified PVCL of similar molecular weight (Fig. 24a). The hydrodynamic radius of PVCL is of the same order of magnitude in aqueous solution and in tetrahydofuran, a good sol-

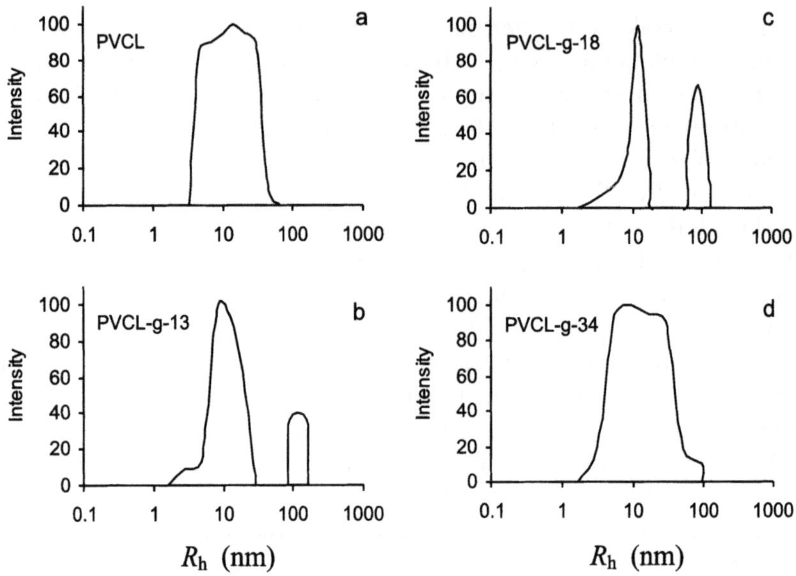

Fig. 24 Size distributions of PVCL and the graft copolymers in water at 20 °C. Homopolymer PVCL (**a**), grafted copolymers PVCL-*g*-13 (**b**), PVCL-*g*-18 (**c**), and PVCL-*g*-34 (**d**). Polymer concentration $10\,g\,L^{-1}$. (Reprinted with permission from Ref. [180] copyright 2005 Elsevier)

vent of PVCL (5–30 nm). Thus, the polymer is dissolved as single chains in water at 20 °C, and this is the situation over the entire concentration range covered.

Two different patterns emerged from the DLS studies of the copolymers, depending on their grafting density. The hydrodynamic radius distribution recorded with solutions of the copolymers of high-grafting density is monomodal ($c = 1-10\,g\,L^{-1}$), with no indication of the presence of larger objects (Fig. 24d). Thus, in aqueous solutions, this highly grafted PVCL sample does not form interpolymeric associates: the number of amphiphilic grafts is high enough to force the polymer chain to adopt a stable intrapolymeric structure. A similar observation was previously reported in a study of PNIPAM grafted with amphiphilic perfluorooctyl groups [252], another type of amphiphilically modified polymers which tend to form unimers rather than interpolymeric associates.

In contrast, the hydrodynamic radius distribution recorded for aqueous solutions of copolymers with a low grafting density is bimodal, with a contribution from small entities of $R_h = 5-30$ nm, assigned to single polymer chains, and a contribution from larger particles of $R_h = 80 = 150$ nm (Fig. 24b, c). The relative importance of the two populations depends on copolymer concentration: the relative amount of the larger particles increases with increasing copolymer concentration.

The self-association of grafted PVCL copolymers at 20 °C was also studied with fluorescence measurements with pyrene (Py) as an extrinsic probe [180]. The variations of the ratio I_1/I_3, a measure of the polarity of the probe environment, were monitored as a function of polymer concentration (Fig. 25). For all copolymer solutions the ratio I_1/I_3 decreased with increasing polymer concentration and eventually levelled off for solutions of copolymer concentration in excess of 0.3–1.0 g L^{-1}, depending on the grafting level. The onset of the decrease in I_1/I_3, which is often taken as an indication of the onset of micellisation, depended on the grafting level of the copolymer: the higher the grafting level, the lower the onset of micellisation (Fig. 25). However, when the data are plotted against the macromonomer concentration it appears that micellisation begins when the $C_{11}EO_{42}$ concentration is more than 4×10^{-3} mmol L^{-1}, for all copolymers. Thus, micellisation of the copolymers is controlled to a large extent by the hydrophobic assembly of the substituents, as noted earlier in the case of various hydrophobically modified polymers [311, 312]. The fact that the minimum $C_{11}EO_{42}$ concentration required for copolymer micellisation is lower by more than 2 orders of magnitude than the critical micelle concentration of the macromonomer is also in agreement with previous observations.

The plots of I_1/I_3 vs. copolymer concentration also reveal differences in the micropolarity of the hydrophobic domains created upon association of the various copolymers in water. A qualitative assessment of this property is given by the I_1/I_3 value determined in the copolymer solutions of highest concentration when the plateau value is attained (Fig. 25). This value depended significantly on the grafting level: the solution of the most densely grafted copolymer yielded the lowest I_1/I_3 value (1.40) and the pure homopolymer the highest. In all cases, this value is higher than the value (1.20) recorded for micellar solutions of the macromonomer. It can be concluded

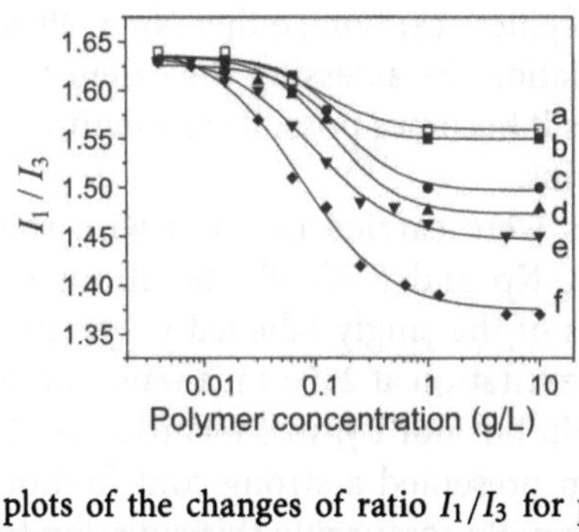

Fig. 25 Semilogarithmic plots of the changes of ratio I_1/I_3 for pyrene in aqueous solutions of homopolymer PVCL and grafted copolymers at 20 °C as a function of polymer concentration. Homopolymer PVCL (*a*), PVCL-*g*-6 (*b*), PVCL-*g*-13 (*c*), PVCL-*g*-16 (*d*), PVCL-*g*-18 (*e*), PVCL-*g*-34 (*f*). (Reprinted with permission from Ref. [180] copyright 2005 Elsevier)

that the assemblies formed with lightly modified PVCL are more polar than the more heavily grafted samples, a possible indication of differences in the hydration of the micellar assemblies.

Fluorescently labelled PVCL and the corresponding graft copolymers were synthesised in order to have molecular-level information of the properties of the thermally responsive polymers (Table 7) [180]. Py- or naphthalene (Np)-containing monomers were copolymerised with VCL to obtain labelled PVCL, while copolymerisation of the fluorescent monomers with VCL and $MAC_{11}EO_{42}$ led to fluorescently labelled graft copolymers. It should be noted, however, that the hydrophobic labels may affect the solution properties of the polymers and, thus, one should be careful when comparing the data obtained from the labelled polymers [179]. In fact, it was noted that, unlike PVCL itself, fluorescently labelled PVCLs tend to form intermolecular and intramolecular associates even in the very low concentration regime, i.e. the labelled homopolymers associate via hydrophobic interactions between the labels of different chains. The labels did not seem to affect as drastically the properties of the graft copolymers. Light scattering measurements confirmed that labelled graft copolymers exist in micellar form and that the sizes of the micelles were of the same order as those of their unlabelled counterparts. Analysis of the photophysical properties of the labels linked to the graft copolymers led the authors to confirm that the micelles of amphiphilically grafted VCL copolymers are *intra*polymeric, while the association of labelled PVCL leads to interpolymeric micelles [179]. This conclusion was reached on the basis of nonradiative energy transfer (NRET) [313] experiments involving mixtures of Py- and Np-labeled graft copolymers.

The experiments are based on the fact that the probability of energy transfer between the two chromophores depends strongly on their separation distance [314]. For the Np–Py pair the characteristic distance of the process is 28.9 Å [315]. Thus, the occurrence of NRET between Np and Py in dilute mixed solutions of polymers carrying either Np or Py signals the existence of interpolymeric association. By assessing the extent of energy transfer under various circumstances it becomes possible to monitor closely the interactions between polymer chains.

NRET experiments were carried out first with solutions of the labelled homopolymers, PVCL-Np and PVCL-Py. Solutions were prepared by mixing aqueous solutions of the singly labelled polymers at room temperature. Their emission upon excitation at 290 nm, a wavelength corresponding to absorption of light by Np but not Py, was recorded at 20 °C immediately after mixing. The spectrum presented a strong contribution from Py emission in addition to Np emission. By comparing this emission to the spectra of the solutions of each polymer of same concentration, a significant decrease of the Np emission was observed together with an increase of the Py emission. The two concurrent effects signal the occurrence of NRET between the two chro-

Table 7 Summary of the reaction conditions in the synthesis of fluorescently labelled thermosensitive polymers [181]

Sample	Composition						
	VCL ($mol\,L^{-1}$)	$MAC_{11}EO_{42}$ ($mmol\,L^{-1}$)	PyAAm ($mmol\,L^{-1}$)	NpAAm ($mmol\,L^{-1}$)	$MAC_{11}EO_{42}$ (mol %)	Label ($mol\,g^{-1}$)	M_w ($g\,mol^{-1}$)
PVCL-Py1	0.90	–	3.0	–	–	3.3×10^{-5}	220 000
PVCL-Py2[a]	0.90	–	3.0	–	–	6.7×10^{-5}	47 000
PVCL-$C_{11}EO_{42}$-Py1	0.90	11.8	3.0	–	1.1	2.8×10^{-5}	290 000
PVCL-$C_{11}EO_{42}$-Py2[a]	0.90	11.8	3.0	–	1.2	4.6×10^{-5}	36 000
PVCL-Np	0.90	–	–	3.0	–	3.2×10^{-5}	210 000
PVCL-$C_{11}EO_{42}$-Np	0.90	11.8	–	3.0	1.1	2.7×10^{-5}	250 000

[a] Polymerised in dimethylformamide

mophores, and hence indicate the existence of interpolymeric associations in the solutions of labelled PVCL kept at room temperature.

The formation of stable particles of PNIPAM [166] or PVCL [182] grafted with PEO was detected earlier and it is expected that the introduction of the amphiphilic PEO–alkyl chains on PVCL, as well, will stabilise the PVCL aggregates upon heating.

The effect of heat on the size of PVCL grafted with $MAC_{11}EO_{42}$ was assessed by light scattering measurements [181]. The grafted PVCL is fully hydrated and dissolved as single chains at low temperature. Upon heating, the hydrodynamic size of the polymer slightly decreases, and, at the cloud point, micelles of individual polymers start to aggregate, causing an increase in their size (Fig. 26, curve a). Above the cloud-point temperature the size of the particles decreases slightly upon further heating. For comparison, the temperature-dependent size of grafted PVCL microgel particles is also presented in Fig. 26 (curve b). Grafted microgel particles exhibit a continuous volume decrease upon heating and finally form collapsed particles at temperatures above 45 °C, Above this temperature, the sizes of microgels and amphiphically grafted polymers are actually very close to each other.

The same phenomenon, i.e. transition from a coil to an aggregated metastable state, was observed for all amphiphilically grafted PVCLs: stable nanoparticles formed in all cases, ranging in size from 50 to 120 nm, depending on the concentration of the solution (Fig. 27) [180, 181]. The size of the particles was hardly affected by grafting density. The size distributions of the particles were rather narrow and monomodal. Homopolymer as well as graft copolymer mesoglobules remained colloidally stable, with no interparticle flocculation over periods of several days, as judged by DLS monitoring of aqueous polymer samples kept at 50 °C.

The stability of the particles was also assessed by dilution experiments. First, particles of a sample of amphiphically grafted PVCL were prepared by

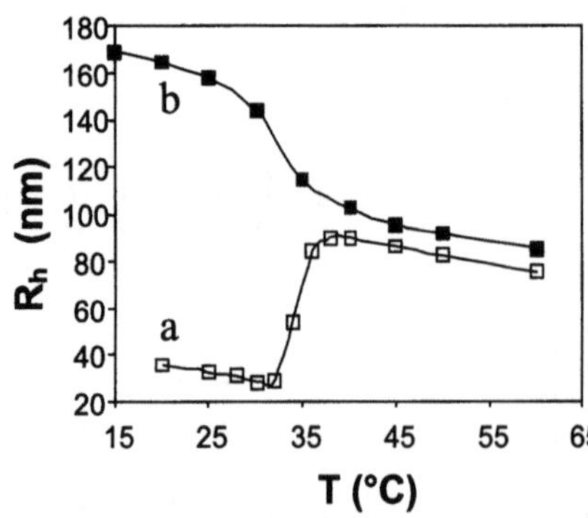

Fig. 26 Hydrodynamic radius of a graft copolymer PVCL-*g*-14 single chain at low temperatures and of the particles formed above the critical solution temperature (*a*). Temperature dependence of the hydrodynamic radius of grafted PVCL microgel E4 (*b*) [181]

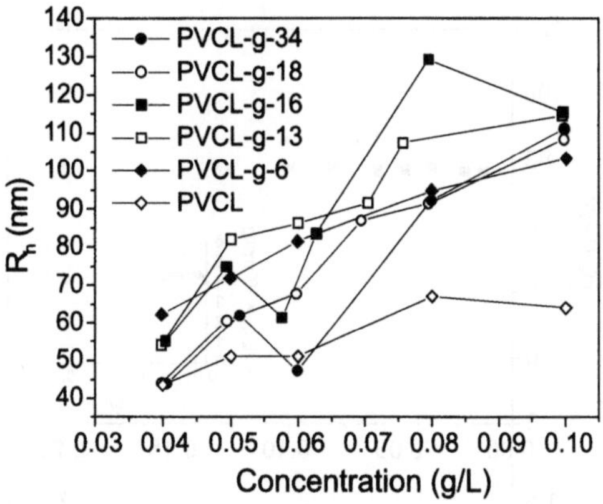

Fig. 27 Apparent hydrodynamic radius (R_h) of the mesoglobules formed by quickly heated aqueous solutions of the homopolymer PVCL or the graft copolymers with varying concentrations. (Reprinted with permission from Ref. [180] copyright 2005 Elsevier)

heating to 50 °C a dilute solution of the copolymer within a light-scattering sample holder [180]. After keeping the sample at 50 °C for 1 h, the first dilution step was conducted by injecting hot water into the solution. The dilution was repeated, once the intensity of the scattered light was stable. The hydrodynamic radius of the particles at 50 °C remained constant, with no sign of particle disintegration upon dilution (Fig. 28, top). Moreover, the intensity of the scattered light from the colloidal sample was directly proportional to the concentration upon dilution (Fig. 28, bottom), providing yet another indication of the stability of the particles.

The apparent molecular weight M_w^{agg} and the radius of gyration R_g of the aggregates were estimated by static light scattering measurements conducted at 50 °C for amphiphically grafted PVCL particle suspensions of three different concentrations, obtained upon dilution with hot water of the most concentrated solution. The second virial coefficient A_2 of the mesoglobules was zero within experimental error (Fig. 28, inset). Accurate estimates of M_w^{agg} are somewhat difficult to obtain because of the uncertainty of the increment of the refractive index, dn/dc. The density of the particles, $\rho = M_w^{agg}/[N_A(4/3)\pi R_h^3]$ indicates that particles are partially hydrated, as in the case of collapsed PNIPAM ($\rho = 0.34\ \text{g cm}^{-3}$) [121]. The aggregation number was estimated to be approximately 1600 ($M_w^{agg} = 4.8 \times 10^8\ \text{g mol}^{-1}$) and the density $0.40\ \text{g cm}^{-3}$, using the dn/dc value of a collapsed PVCL microgel [316]. The shape of the particles was estimated from the ratio R_g/R_h, which ranged from 0.72 to 0.75, indicating that the particles adopt a homogenous spherical geometry even after being subjected to dilution. Qualitatively similar observations were also made for mesoglobules formed by the homopolymer PVCL [147].

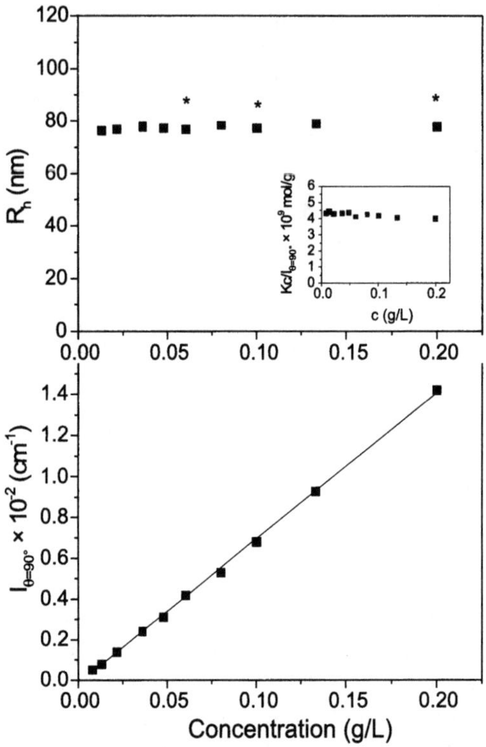

Fig. 28 *Top*: Effect of dilution on R_h and $Kc/I_{\theta=90°}$ of the mesoglobules (■) formed upon heating of 0.2 g L^{-1} aqueous solution of PVCL-*g*-18. *Bottom*: Scattering intensity at 90° from polymer solution of different concentration. (Reprinted with permission from Ref. [180] copyright 2005 Elsevier)

In the mesoglobular phase, at temperatures above the LCST, the polymers are entangled and stuck together via hydrophobic forces [180]. These forces are highly temperature dependent and become ineffective as the polymer solution is cooled below the LCST. Consequently, the particles disintegrate upon cooling. The physical network that is present at elevated temperatures can, however, be strengthened by solubilising within the particles compounds, such as polyphenols, which are able to interact with several PVCL chains via hydrogen-bond formation between the carbonyl groups of the lactam ring and the hydroxyl units of the phenols (Fig. 29). These specific interactions have been used to prepare hydrogels via physical cross-linking of linear PVCL with polyphenols, such as pyrocatechol and phloroglucinol [151, 317, 318]. The same cross-linking mechanism was tested with the colloidal PVCL particles in an attempt to fix permanently the mesoglobule structure, so that the particles preserve their integrity upon cooling.

The effect of added phenols on the size of heat-induced particles was studied via DLS in the case of PVCL and amphiphically grafted PVCL par-

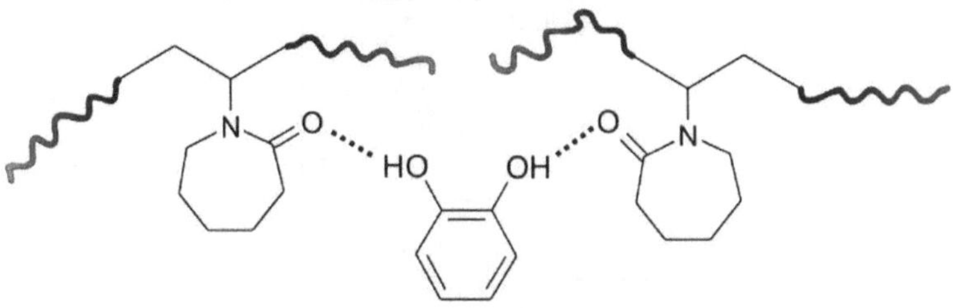

Fig. 29 Formation of physical PVCL network via hydrogen bonding [181]

ticles [180]. Cross-linking was performed by injecting a hot aqueous 1,2-benzenediol solution into a colloidal polymer dispersion placed in a light-scattering cell holder at 50 °C. Addition of the phenol at 50 °C did not affect the size of the PVCL particles, but it triggered a slight decrease of the size of the graft PVCL particles, from $R_h = 77$ to 64 nm. The decrease in the size of the graft copolymer particles may be a result of cross-linking of a less dense structure of the PEO-containing particle compared with that of a PVCL mesoglobule. The treated solutions were allowed to cool down to 20 °C. Their hydrodynamic radii did not change, which indicates that physical cross-linking occurred within the particles. In the absence of the cross-linker, the particles disintegrated immediately as the temperature was lowered below 32 °C. Cross-linked particles were monitored at 20 °C for several weeks by DLS and no flocculation was detected.

4 Poly(vinyl methyl ether)

Aqueous solutions of PVME (Fig. 30) are known to cloud in a temperature window ranging from 32 to 40 °C, depending on molar mass and concentration [154]. The phase diagram of PVME in water is bimodal, presenting two critical points: the point of lower polymer concentration corresponds to type I behaviour, according to the phenomenological classification of Berghmans, whereas the point of higher concentration is a transition of type II (Fig. 31) [161–164].

In contrast to PNIPAM, direct observation of the coil–globule transition of a single PVME chain has not been reported. However, a number of reports

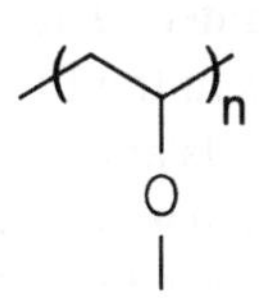

Fig. 30 Poly(vinyl methyl ether) (*PVME*)

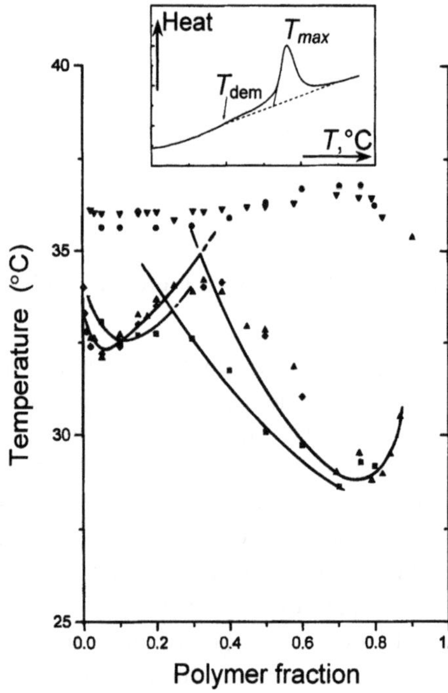

Fig. 31 Phase diagram of type III for PVME in water with two different molar masses (*hand-drawn curves*). T_{dem}: $M_w = 28\,000\ g\,mol^{-1}$ (■) and $M_w = 147\,000\ g\,mol^{-1}$ (▲). Peak temperature T_{max}: $M_w = 28\,000\ g\,mol^{-1}$ (•) and $M_w = 147\,000\ g\,mol^{-1}$ (▼). Cloud point for $M_w = 147\,000\ g\,mol^{-1}$ (♦). Scanning rate 0.1 °C min^{-1}. (Reprinted with permission from Ref. [161] copyright 1997 American Chemical Society)

exist on the properties of aqueous PVME solutions [153–156, 163, 319–325] and of radiation-cross-linked PVME hydrogels [162, 163, 324, 326–330].

IR spectroscopy was used by Maeda [320] to study polymer–solvent interactions during the phase transitions of PVME in water. This technique provides detailed information on the hydration states of individual functional groups of PVME both below and above the cloud point. In solutions below the phase transition, the ether oxygens of the polymer units form hydrogen bonds with water molecules. This enthalpic contribution to the free energy of solution overrides the unfavourable decrease in entropy due to the formation of a layer of organised water around the hydrophobic moieties of the polymer chain. With increasing temperature, from about 33 to 40 °C, the hydrogen bonds between water molecules and the polymer ether oxygens are broken and, at the same time, nonpolar groups are dehydrated. These changes in water–polymer interactions are reflected by changes in the IR bands attributed to the vibration modes of the PVME backbone bonds and of the methoxy groups. Careful analysis of these spectral features led Maeda to conclude that most of the methyl groups of PVME are dehydrated above the transition temperature, whereas the ether groups are only partially dehydrated.

The temperature-induced phase transition in PVME–D_2O solutions within a broad concentration range ($c = 1–300\,g\,L^{-1}$) and PVME gels of various cross-linking densities was monitored by ^1H NMR spectroscopy by Spěváček et al. [321–323], to investigate changes in their dynamic structure. By dynamic experiments, it was possible to observe the formation of compact globulelike structures for both linear and cross-linked systems. In dilute PVME solutions, the transition detected by NMR is virtually discontinuous, taking place over a narrow temperature range. In semidilute and concentrated solution, as well as in concentrated swollen networks, the transition begins at lower temperatures and occurs over a temperature range several degrees wide.

Further studies confirmed the formation of compact globular structures in PVME–D_2O solutions kept at elevated temperatures [324]. In such solutions most methyl groups of PVME are dehydrated. The globulelike structures are more compact in heated dilute solutions, compared with those formed in semidilute or concentrated solutions, for which the globules may still contain a certain amount of water. Dehydration of PVME chains occurs rapidly in dilute solutions. In contrast, in semidilute and concentrated solutions, the originally bound water is squeezed out of globular-like structures very slowly. With time, the character of these globules changes from spongelike to rather compact objects [324].

Temperature- and composition-dependent small-angle neutron scattering measurements were carried out recently with homogeneous mixtures of PVME and D_2O [331]. The experimental data yielded the values of the second-order compositional derivative of the Gibbs energy, the Ornstein–Zernike correlation length, and the LCST spinodal temperatures. The miscibility behaviour, predicted using an extended Flory–Huggins interaction function, was in good qualitative agreement with the experimental data. On the basis of PVME molar mass values in D_2O and toluene, it was concluded that, in dilute and concentrated solutions at 20 °C, D_2O and PVME do not form stable molecular complexes. The energy needed to induce compositional fluctuations was found to be only about $0.3\,k_BT$, even at the lowest temperature investigated. In the temperature interval probed, specific interactions between water or D_2O and PVME are too weak to bring about the formation of stable complexes.

Light scattering and electron microscopy studies of aqueous PVME solutions and PVME microgels were carried out by Arndt et al. [329, 330]. They noted that the M_w of PVME in water was always higher (up to 20 times) than its value ($M_w = 46\,000\,g\,mol^{-1}$) determined in organic solvent (butanone), even for dilute aqueous PVME solutions well below the phase-separation temperature [330]. Moreover the molar masses of the polymer in water depended on solution preparation conditions. The authors concluded that PVME does not exist as isolated chains in water, but forms loose aggregates ($R_h = 200–220$ nm) which decrease in size as the solution temperature passes from 25 to 36 °C. Above 36 °C, the shape of the PVME aggregates changes. Mesoglobules of PVME are stable with increasing temperatures, at least up to 60 °C. One should point out that the studies were carried out with a commercial PVME, which was not assessed for the possible presence of residual charges on the PVME chains. As discussed in Sect. 2, in the case of PNIPAM, a minute number of charges along a chain may enhance the colloidal stability of mesoglobules [330].

However, recent measurements on aqueous solutions of two PVME samples ($M_w = 12\,800$ and $19\,600\,\text{g mol}^{-1}$) known to be devoid of charged groups confirmed the formation of stable mesoglobules at elevated temperature [147]. The R_h of the loose PVME aggregates formed in cold solutions ($0.02–0.25\,\text{g L}^{-1}$) was shown to decrease upon heating, up to the cloud-point temperature (approximately 35–37 °C). Above 40 °C, colloidally stable mesoglobules, of monomodal and extremely narrow size distribution formed (Fig. 32) [147]. From analysis of a Kratky plot [332] obtained for PVME mesoglobules in water at 40 °C, it was concluded that the mesoglobules adopt the conformation of a soft sphere [330]. The polymer density is not uniform throughout such spheres: it decreases with increasing distance from the centre. The scattering function recorded from solutions kept below their phase-transition temperature was indicative of the presence of branched, starlike loose PVME associates [330].

Thermosensitive microgel particles ($R_h = 300–500$ nm) were synthesised by electron beam irradiation of dilute aqueous PVME solutions [330]. It was noted that when the irradiation of the PVME solution ($4.0\,\text{g L}^{-1}$) was

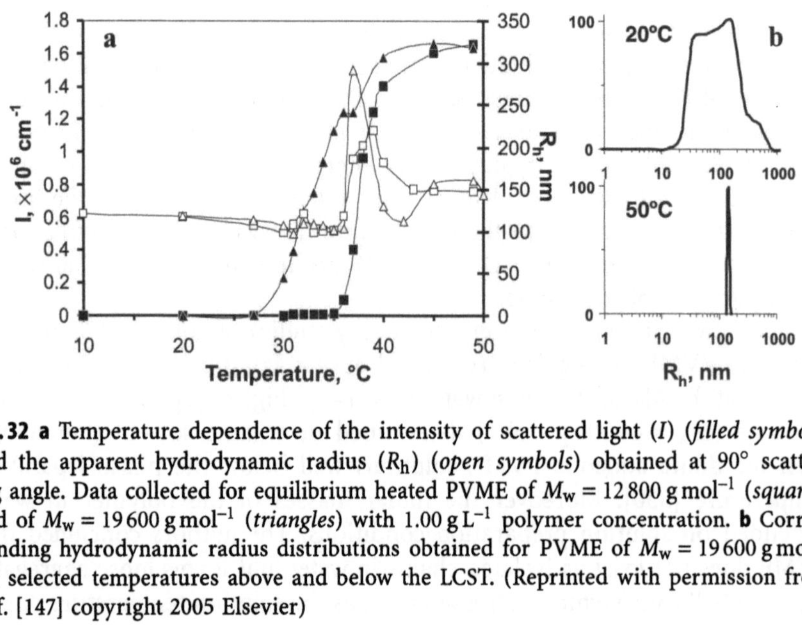

Fig. 32 a Temperature dependence of the intensity of scattered light (I) (*filled symbols*) and the apparent hydrodynamic radius (R_h) (*open symbols*) obtained at 90° scattering angle. Data collected for equilibrium heated PVME of $M_w = 12\,800\,\text{g mol}^{-1}$ (*squares*) and of $M_w = 19\,600\,\text{g mol}^{-1}$ (*triangles*) with $1.00\,\text{g L}^{-1}$ polymer concentration. **b** Corresponding hydrodynamic radius distributions obtained for PVME of $M_w = 19\,600\,\text{g mol}^{-1}$ for selected temperatures above and below the LCST. (Reprinted with permission from Ref. [147] copyright 2005 Elsevier)

performed at temperatures above the solution phase-transition temperature (60 °C), the mesoglobular structure was preserved in the resulting microgel. Above the phase-transition temperature, PVME microgel particles have a spongelike structure. The concentration employed in these studies is significantly higher than the highest possible concentration, which allows the formation of colloidally stable PNIPAM and PVCL solutions [147], underlining the high stability of PVME mesoglobules.

5
Mechanisms of Colloidal Stability of Neutral Water-Soluble Amphiphilic Polymers

5.1
Colloidal Stability and Mesoglobules

When polymer solutions encounter situations for which the thermodynamic quality of the solvent becomes poor, individual chains undergo a coil-to-globule collapse, the globules associate immediately, and macroscopic phase separation seems unavoidable. However, it has been reported that a number of polymers in water or in organic solvents form equilibrium globules, i.e. single-chain globules that remain in solution without immediate association and precipitation.

For example, in the case of PS and applying the Smoluchowski equation [333], it is possible to estimate the precipitation time, t_{pr}, of globules of radius R and translation diffusion coefficient D in solutions of polymer concentration c_p (the number of chains per unit volume) [334]. Assuming a standard diffusion-limited aggregation process, two globules merge every time they collide in the course of Brownian motion. Thus, one can write Eq. 2:

$$t_{pr} \sim \frac{1}{4\pi D R c_p} \sim \frac{3}{2c_p} \frac{\eta_0}{k_B T}, \qquad (2)$$

where η_0 is the solvent viscosity and T is the temperature. Applying this equation to the experimental conditions (PS in cyclohexane subjected to cool-quenching from 35 to 28 °C) used by Chu et al. [83], one finds $t_{pr} \sim 0.1$ s, whereas experimentally it was observed that aggregation starts only after about 10 min [83]. The time dependence of R_h for PS in solutions kept at 28 °C was described as a sum of two exponential functions of $\exp(-t/\tau)$ type. The relaxation time required for a coil to adopt a crumpled globule state was determined experimentally to be 357 s. The relaxation time for adopting the compact equilibrium globule was 323 s. Thus, one can estimate that globules in a poor solvent may collide several thousand times before merging takes place [334]. These experiments on PS collapse in cyclohexane unambiguously demonstrate that the lifetime of individual PS globules, without significant precipitation, is much longer than the time required for two individual globules to collide [83]. Similar observations were reported in the case of single-chain globules of poly(methyl methacrylate) in isoamyl acetate which were observed to form aggregates 10–60 min after the coil-to-globule transition [97]. Mesoglobules of poly(ε-caprolactone) ($M_w = 20\,000$ g mol^{-1}) in tetrahydrofuran have also been reported [267]. PNIPAM ($M_w > 10^6$ g mol^{-1}) globules in water were reported to be stable for up to 3 days in solutions kept at a temperature just above their LCST [120].

Other relevant examples include poly(1,2-dimethoxy ethylene) (M_w = 18 000, 25 000, and 80 000 g mol^{-1}) which undergoes a coil–globule transition above 60 °C in aqueous solutions of concentration lower than 0.06 g L^{-1} [335]. No precipitation was detected in solutions heated up to 100 °C, though the stability of the globules against aggregation and precipitation was not monitored as a function of time. PNIPMAM and poly(*N*-propyl methacrylamide) globules in water ($c < 0.2$ g L^{-1}) also exhibit some degree of resistance against precipitation when heated above their respective phase-transition temperature [336].

Typically, in the dilute regime, aqueous solutions of thermosensitive water-soluble polymers of moderate molar masses become "cloudy", "milky", or "opaque" when heated above their cloud point. This observation is usually taken as an indication of microscopic phase separation via a metastable phase that will eventually lead to precipitation. However, we have presented in the preceding sections instances for which heating aqueous solutions of thermosensitive polymers leads neither to stable single-chain globules nor to macroscopic phase separation. Instead, stable aggregates form upon heating. The process of intermolecular aggregation stops when particles of a certain size are formed. These aggregates, which arise in dilute solutions ($c < 1.0$ mg mL), may have sizes on the order of 50–300 nm [141–147, 330]. Such aggregates are called mesoglobules, to distinguish them from single-chain globules.

The observation of multimolecular aggregates does not rule out the possibility that thermosensitive polymers of high molar mass adopt a single-chain globular conformation, colloidally stable above the cloud point in extremely dilute solutions. Comparing the various systems, one may conclude that under certain external conditions (heating rate, polymer concentration, etc.), long polymer chains form single-molecule globules, whereas shorter chains associate forming spherical monodisperse mesoglobules of the same morphology, i.e. with the same fractal dimension. Thus, for PNIPAM mesoglobules and globules, the fractal dimension was found to be 2.7 (Fig. 33) [147]. The R_g/R_h ratio determined for various mesoglobular systems fluctuates around 0.77, a value typical for homogeneous nondraining spheres. The density of the globular particles is on the order of 0.35–0.40, indicating that the mesoglobules still contain solvent [87, 119–121, 147]. This value coincides with theoretically predicted 0.40 g cm^{-3} for a space-filling model [337]. Taking into

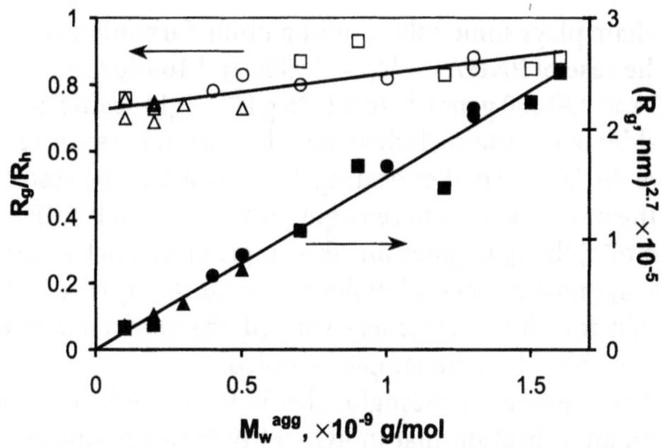

Fig. 33 Scaling of the molar mass of PNIPAM mesoglobules (M_w^{agg}) vs. their radius of gyration (R_g) with fractal dimensionality 2.7 (*filled symbols*) and the shape factor R_g/R_h (*open symbols*). The conditions at which mesoglobules were formed correspond to those in Table 2: $M_w = 27\,300\ \text{g mol}^{-1}$, non-equilibrium heated (*circles*); $M_w = 160\,000\ \text{g mol}^{-1}$, nonequilibrium heated (*triangles*); $M_w = 160\,000\ \text{g mol}^{-1}$, equilibrium heated (*squares*). (Reprinted with permission from Ref. [147] copyright 2005 Elsevier)

account the experimental static scattering factor, the density, and the narrow size distribution of the mesoglobules, one can assume that they can be seen as homogeneous nondraining spheres with rough surfaces.

To account for the unexpected colloidal stability of mesoglobules, it was suggested in several publications that the solutions underwent spinodal decomposition above their LCST and that they did not undergo macroscopic phase separation owing to the low probability of Brownian collisions in dilute solutions [141–145]. A polymer in a poor solvent tends to find optimal interaction energy conditions via intermolecular association among polymer chains, which results in a decrease of the chain translational entropy. When these two energy terms are compensated, the system reaches a metastable state characterised by the formation of polymeric mesoglobules. As noted earlier, in some of the first systems for which this stability against precipitation was observed, the polymer chains (PNIPAM or PVME) may have had charged chain termini that may have contributed to the colloidal stability of the mesoglobules [141–145, 330]. More recently, mesoglobules of PNIPAM and PVCL ($M_w \sim 10^4$–$10^5\ \text{g mol}^{-1}$) synthesised under conditions leading to electroneutral chains in the absence of any additives were observed above their respective solution LCSTs, pointing to the occurrence of a stabilisation mechanism other than the classic electrostatic colloidal stability [146, 147].

Factors affecting the size of the mesoglobules include polymer concentration, i.e. the size increases with solution concentration and, more importantly, heating rate (Fig. 34). Thus, a fast increase in temperature (nonequilibrium heating) leads to mesoglobules of smaller size than those formed upon slow heating through the sample LCST [141–145, 147]. The chemical composition of the polymer chain plays some role in mesoglobule formation and stability. For example, in the case of PNIPAM ($M_w = 27\,300$ and $160\,000$ g mol^{-1}) and PVCL ($M_w = 30\,000$ and $330\,000$ g mol^{-1}) ($c < 0.25$ g L^{-1}) light scattering data suggest that a metastable state is attained when the solution temperature reaches a value 5–10 °C above the LCST. Further heating does not affect the system, at least as judged from the intensity of scattered light. In solutions of PVME ($M_w = 12\,800$ and $19\,600$ g mol^{-1}) the aggregates shrink to a certain extent before forming colloidally stable aggregates/mesoglobules above 40–45 °C (Fig. 32). It is worth mentioning that for all the polymers studied the conformational change is completely reversible when the sample is cooled.

Another characteristic of mesoglobules is their stability against dilution with hot water, an indication that individual polymer chains are trapped inside the mesoglobules [144, 145, 147]. Dilution of the samples with hot water does not induce any changes in the apparent R_h of the aggregates (Fig. 34). Moreover, the inset in Fig. 34 showing a $Kc/I_{\theta=90°}$ vs. c plot clearly demonstrates that, within experimental error, the second virial coefficient, A_2, of the mesoglobules is zero. Therefore, the molar masses of the aggregates can be determined at each concentration separately. Accordingly, the R_h value extrapolated to zero scattering angle is a real value, hardly affected by either intermolecular thermodynamic or hydrodynamic interactions.

The stability of the dispersions upon dilution at 50 °C and zero value of the osmotic second virial coefficient suggests that the surface of the particles at temperatures above the LCST may possess some hydrophilic character: the macromolecules self-organise and build up particles with hydrated po-

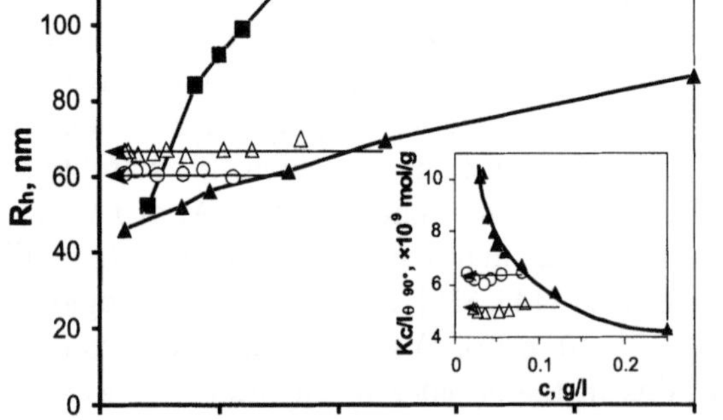

Fig. 34 The effects of the heating rate and dilution with hot water of mesoglobules of PNIPAM 160 000 g mol^{-1} on R_h and $Kc/I_{\theta=90°}$: equilibrium heated (■), nonequilibrium heated (▲), dilution of 0.12 g L^{-1} solution (△), and dilution of 0.08 g L^{-1} solution (○). All the solutions are at 50 °C. $dn/dc = 0.20$ cm^3 g^{-1}. (Reprinted with permission from Ref. [147] copyright 2005 Elsevier)

lar groups turned towards the surrounding aqueous phase. Such amphiphilic polymers are predicted to adopt a cylindrical conformation, owing to their large surface-to-volume ratio [139]. However, the existence of cylindrical mesoglobules has not been observed [147]. Mesoglobules of PNIPAM, PVCL, and PVME are spherical, similar to single-chain globules [119–121].

What could be the reason for this remarkable slowing down of aggregate association? Why are mesoglobules not as "sticky" as anticipated, given the thermodynamically poor conditions to which they are subjected?

The competition between hydrophobic and hydrophilic interactions with respect to the phase behaviour of thermosensitive polymers in water has been examined theoretically, using both mean-field theory and computer simulations [139, 338, 339]. The following scenario emerges. When an amphiphilic macromolecule collapses, the hydrophobic backbone folds, whereas the hydrophilic groups turn outside towards the water molecules. The macromolecule adopts an intermediate necklacelike conformation where hydrophilic groups surround single pearls of hydrophobic groups. Stretched chain sections interconnect pearls. Reducing the thermodynamic quality of the solvent results in an increase of pearl size and a decrease of the number of pearls. Finally, the pearls coalesce and form a cylindrical particle.

This mechanism may account for the stability, in the absence of any external stabilising agent, of amphiphilic homopolymers in the fully collapsed/globular state. The total free energy of a collapsed macromolecule includes a surface energy contribution in addition to the bulk free energy. Obviously, to form a stable particle, the outer shell of the particle should be hydrophilic enough.

5.2 Partial Vitrification

Taken together, the experimental observations reported in the previous sections suggest that the formation of colloidally stable particles heated above their phase-transition temperature may be a universal phenomenon, taking place not only in aqueous polymer solutions, but also in solutions of polymers that can undergo a coil–globule transition in organic solvents.

When a polymer solution that shows a UCST is cooled down, phase separation results in a polymer-rich phase of high density. Actually, the density of the polymer-rich phase is not as high as that of the polymer in bulk, owing to the plasticising effect of the solvent. At low temperatures, the dynamics of the polymer in the polymer-rich phase depends on the free energy profile of the system, represented in Fig. 35. If the temperature is low enough, the polymer-rich phase adopts a glassy state and various metastable morphologies may coexist; since the free-energy barriers between metastable states are high, morphological changes are restricted. In the mean-field limit, the metastable states have an infinitely long lifetime [340, 341]. Following this

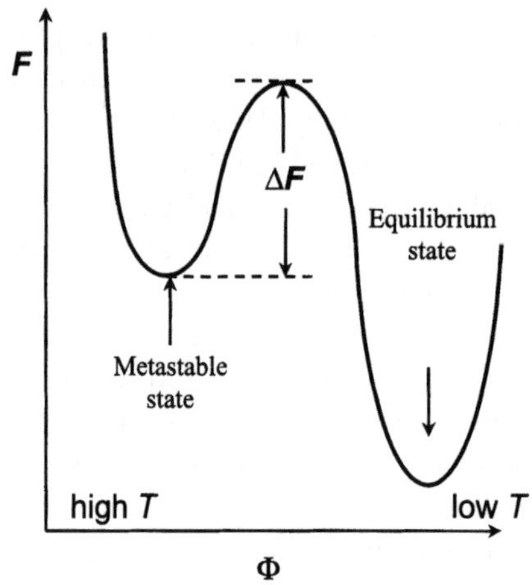

Fig. 35 An illustration of a metastable state in a plot of free energy (F) vs. the order parameter (Φ), where ΔF is the activation barrier. (Schematically reproduced from Ref. [105])

concept, polymer globules formed in dilute solutions are expected to "freeze" and the aggregation to stop.

Keller et al. [105–107, 342–344] presented a thermodynamic approach to the phase separation of polymers showing a UCST. They discussed the thermodynamic stability of the polymer–solvent system and discussed the various metastable states adopted under thermodynamically poor conditions by the full range of polymer–solvent compositions. Figure 36 schematically depicts the multiple metastable morphologies possible in a polymer–solvent system below the phase-separation boundary, depending on the polymer content. In a single-component liquid (i.e. polymer melt) vitrification occurs at the glass-transition temperature, T_g, of the dry polymer. The T_g decreases upon dilution and the $T_g(c)$ curve may intersect with the phase-separation curve at a given point (often called the Berghmans point) (Fig. 36) [345]. Actually, the T_g value cannot be lower than the temperature corresponding to the Berghmans point; T_g is invariant for polymer concentrations (Berghmans plateau) lower than the concentration of polymeric material in the collapsed state and further compositional changes are not possible. At temperatures below the Berghmans point, the polymer becomes glassy, "frozen in", and, consequently, phase morphology is preserved.

In dilute polymer solutions, collapse of the polymer upon cooling, indicated by the arrow in Fig. 36, results in an increase in the density of repeating units within the globule. If the temperature of the transition is at or lower than the Berghmans point, i.e. lower than the T_g of the polymer at the concentration corresponding to the globule density, the polymer vitrifies. In dilute solutions, the vitrified polymer-rich phase will be in the form of small dispersed spheres representing a morphologically metastable state [105, 342].

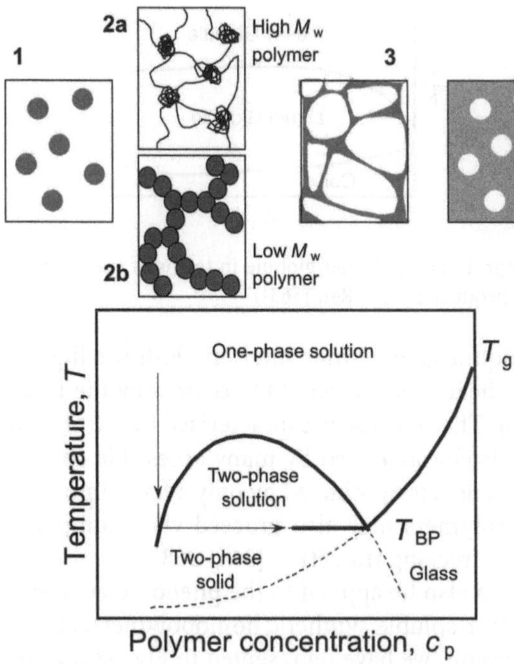

Fig. 36 Phase diagram for a polymer with an upper critical solution temperature plotted together with the vitrification curve schematically illustrating the work of Berghmans and Keller. Pictures represent morphological structures formed below the phase boundary: dispersed glassy phase in a solvated matrix (*1*), molecularly connected phase (glassy spheres–polymer rich phase) for high molar mass polymers (*2a*), dispersed glassy phase connected through adhesive contacts for low molar mass polymers (*2b*), polymer phase-connected bicontinuous phase (*3*), and dispersed solvated phase ("globules" of solvent) in a glassy polymeric matrix (*4*). (Schematically reproduced from Ref. [343])

Interactions of such glassy polymeric particles should resemble the collisions of hard spheres. Phase diagrams of the type shown in Fig. 36 have been obtained for various polymer–organic solvent mixtures [85, 94, 345–353].

The glass transition in homopolymer globules has been modelled by means of Monte Carlo simulations and molecular dynamics simulations [354–362]. The interactions between globules are represented by a negative (attractive) second virial coefficient, A_2, and the final state of the globule is defined by a positive (repulsive) third virial coefficient, A_3 [54]. The formation of a liquid globule is expected to occur in a solvent just below its Θ-temperature, whereas the glassy state is adopted in a poor solvent far beyond the Θ-temperature (Fig. 37). An analytical mean-field theory of the collapse of a linear homopolymer, developed by Rostiashvili et al. [363], predicts that beyond a certain density, a liquid globule freezes and forms a glass. The transition between a liquid globule and a glassy globule is predicted to be a first-order transition and therefore an abrupt one [356–358, 363]. Since

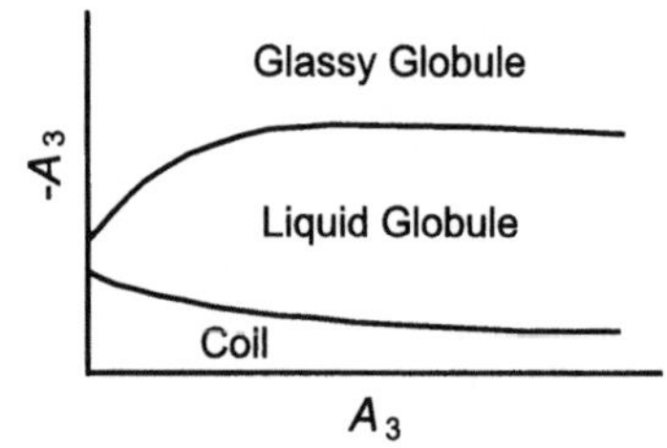

Fig. 37 The phase diagram of a polymer globule in terms of the virial coefficients A_2 and A_3. (Schematically reproduced from Ref. [363])

the collapse of a liquid globule may interfere with the liquid–glassy globule transition, a latent heat was predicted to accompany the liquid–frozen globule transition [363]. The freezing is characterised by a transition between two phases: one phase is characterised by many accessible conformations, while the other, the frozen one, is dominated by only a few conformations. Note that the collapse of copolymers may also proceed via a compact globule, which undergoes a further freezing transition [364–368].

Can this approach also be applied to the phenomena taking place in aqueous solutions of water-soluble synthetic homopolymers, such as PNIPAM? To illustrate the discussion, we have represented in Fig. 38 the concentration dependence of the glass-transition temperature, T_g, of PNIPAM together with the demixing temperature, T_{dem}, of its aqueous solutions [159, 160]. The width of the glass transition is an important parameter, since phenomena related to a putative partial vitrification during the phase separation depend on the upper limit temperature of the glass transition rather than on the temperature corresponding to the mean midpoint of the transition. The inset in Fig. 38 represents the temperature dependence of the apparent specific heat capacity during nonisothermal demixing. The phase separation process is accompanied by an endothermic heat effect (Fig. 38, zone B). When PNIPAM is heated up, just above the demixing boundary where the depth of quenching is small, the polymer-rich phase is liquidlike (zone B) [363]. Further heating of PNIPAM solutions results in a drop in the apparent specific heat capacity below the extrapolated baseline, which was suggested to correspond to the partial vitrification of the PNIPAM-rich phase (zone C, arrow). The temperature at which the polymer-rich phase vitrifies is affected by the heating rate.

Vitrification of the polymer-rich phase during phase separation was demonstrated also in the case of a 70/30 PNIPAM ($M_w = 74\,000$)–water mixture heated to 60 °C (zone C) and subsequently cooled by fast quenching with liquid nitrogen to prevent remixing of the vitrified polymer-rich phase and the surrounding water during cooling [160]. Reheating of the quenched phase led first to melting of the ice. (i.e. water-rich phase), followed by devitrification of the polymer-rich phase at 32 °C (width of $T_g = 14$ °C). This temperature is almost identical to the temperature for which the apparent specific heat capacity drops below the C_p baseline, but it is different from the

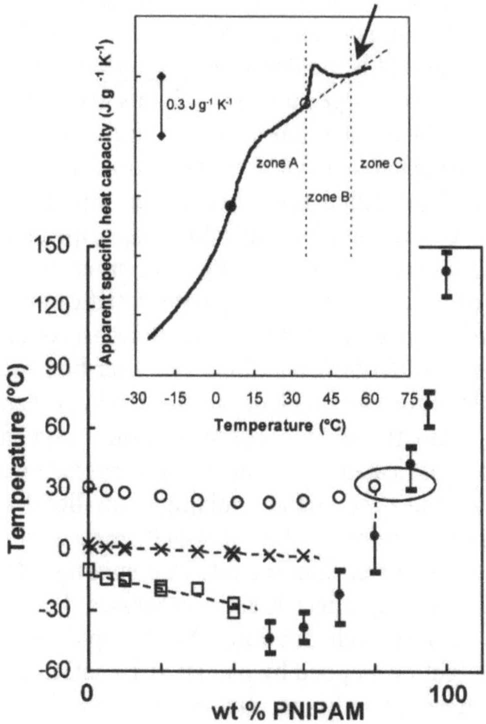

Fig. 38 State diagram of PNIPAM with $M_w = 74\,000\ \mathrm{g\,mol^{-1}}$ in water: demixing curve (○), T_g–composition curve (•, width *error bars*), melting T_m (×) and crystallization T_{cr} (□) temperature of water. Possible interference of partial vitrification during phase separation is indicated by the *oval*. *Inset*: Temperature dependence of apparent specific heat capacity during nonisothermal demixing for 80/20 PNIPAM 74000–water composition: T_{dem} (○) and T_g (•). The *dashed line* is the extrapolated experimental base line. Temperature regions: homogeneous (*zone A*), heterogeneous without interference of vitrification (*zone B*), and heterogeneous with partial vitrification of polymer-rich phase (*zone C*). Partial vitrification of the PNIPAM-rich phase is indicated by an *arrow*. (Adapted with permission from Ref. [160] copyright 2004 American Chemical Society)

T_g (−22 °C) of a homogeneous 70/30 PNIPAM–water mixture. Observation of samples by scanning electron microscopy and optical microscopy revealed that the morphology of the polymer-rich phase is preserved only if the polymer solutions are brought to zone C. Polymer solutions heated to zone B undergo demixing upon quench-cooling [160]. Aqueous solutions of PVCL, PNIPMAM, and PNIPMA exhibit similar behaviour [157, 158, 369, 370].

Can vitrification of PVCL, PNIPAM, and PNIPMA also take place within mesoglobules generated upon heating aqueous solutions of the respective polymers above their LCST? A mesoglobule formed by a large number of short chains is expected to have a lower T_g than a single chain of the same size, since as a rule the T_g of a polymer increases with molar mass [334]. The physical state of the polymer may also depend on the size of the mesoglobule. A macromolecule in its fully collapsed conformation resembles a hard nondraining sphere and its size $R \sim N^{1/3}$, where N is the degree of polymerisation. At the same temperature, an aggregate of a large number of chains may still resemble a sphere, but the polymer may be in the molten state. In this situation, each short chain adopts a more swollen Gaussian coil conformation, for which $R \sim N^{1/2}$.

If the mesoglobules formed upon heating an aqueous polymer solution undergo partial vitrification, this vitrification has some impact during the reverse globule–coil transition upon cooling. A number of experimental results may be interpreted as an indication of partial vitrification. For example, the hysteresis of a cycle coil–globule–coil transition observed in studies by light scattering of aqueous PNIPAM solutions was ascribed to the formation of a new molten-globule state arising from additional hydrogen bonds in the collapsed state [371]. Also, the globule–coil transition is expected to be retarded, since when a sample is heated above the onset of partial vitrification, the specific heat capacity value after partial remixing is smaller than the initial value of the homogeneous solution. In fact, TM-DSC measurements revealed that the rate of remixing is slower than the rate of demixing [160].

Partial vitrification may affect kinetic processes during the coil–globule transition. Thus, at very high dilution, macroscopic phase separation well above the LCST might be stopped by partial vitrification of the polymer-rich phase. At this point we can only speculate whether vitrification interferes with the coil–globule transition or not. This problem is open for discussion and needs experimental confirmation.

Finally, a word about aqueous PVME solutions. In this case as well, stepwise quasi-isothermal measurements through the phase transition revealed that the remixing process (cooling) was slower than demixing, as also seen in the case of aqueous PVCL and PNIPAM solutions [164]. However, the T_g of PVME in the bulk is very low ($T_g = -19\,°C$ [163]) and does not allow overlap of the vitrification and coil–globule transition processes [163, 164]. In fact, the drop in the apparent specific heat capacity below the extrapolated baseline in the quasi-isothermal measurements of PVME solutions was insignificant, compared with the situation noted for PVCL and PNIPAM solutions. The heat capacity of the PVME polymer-rich phase above the LCST does not differ much from that of water, suggesting a liquid state of PVME. One has to conclude that stabilisation mechanisms other than partial vitrification must be at work in the case of PVME mesoglobules.

5.3 Viscoelastic Effect

We examine next how the viscoelastic effect, recently discussed in the literature [190, 267–270], may also account for the resistance of globules and mesoglobules against precipitation.

As already mentioned, the stability of polymeric dispersions can be discussed in terms of two characteristic times: (1) the time between collisions of two globules and (2) the time of reptation of a chain close to the globule surface. A collision of two particles is ineffective, i.e. elastic, if the characteristic time of the collision, i.e. the time of contact, t_c, is less than the time, t_e, required to establish a permanent entanglement between two colliding globules via chain reptation. t_e is expected to be large for entangled chains in the collapsed state and, consequently, most collisions are ineffective. The globules resemble elastic bodies that collide many times before coalescence occurs [372, 373]. This approach not only rationalises the stability of mesoglobules, but also explains why mesoglobules prepared by cool/heat quenching are smaller than those cooled/heated slowly. Quenching results in quick intrachain contraction, preventing extensive interchain association. Therefore, the size of mesoglobules and, therefore, the degree of intermolecular aggregation depend on the depth of quenching and the rate of cooling/heating. On the basis of their own investigations of the collapse of poly(methyl methacrylate) in isoamyl acetate, Baysal and Karasz [97] concluded that the total collapse can occur only well below the Θ-temperature if the depth of quenching (i.e. the shock cooling depth, ΔT) is very deep, which provides stability of globules.

To describe the merging of two globules, Chuang et al. [334] introduced an additional "entanglement force" operational on the prereptational time scale. Merging of two globules proceeds via reptation and takes about half the time required for a globule to diffuse a distance equal to its diameter, i.e. $t = R^2/D$. However, globules approaching each other are dense and highly entangled; thus, intraglobular mobility is limited, resulting in topological restrictions on a time scale shorter than that of reptation. No entropy can be gained owing to topological prohibition of the majority of conformations, consequently, merging is entropically unfavourable. Before reptation has time to occur, the globules experience an entropic counterforce or topological repulsion, called "entanglement force" [334]. The work needed to overcome the entanglement force was estimated to be on the order of approximately $10\,k_B T$. In summary, the model developed by Chuang et al. predicts that numerous collisions of globules must happen before effective entanglement may occur. The question of how deeply the globules have to interpenetrate before coalescence becomes irreversible still remains to be answered [334].

All the aqueous polymers known to form stable mesoglobules at elevated temperature are amphiphiles. We noted earlier that a hydrophilic envelope might protect the mesoglobules against precipitation, via a colloidal stabilisation mechanism. The hydrophobic groups as well may contribute to mesoglobule stability. Hydrophobic association inside mesoglobules may enhances the stability of mesoglobules by increasing the entanglement time t_e [268, 281]. In dilute copolymer solutions at elevated temperatures, macromolecules containing hydrophobic comonomers associate and collapse adopting a metastable state, without macroscopic precipitation [268]. Hydrophobic comonomers also lead to the formation of smaller mesoglobules, as demonstrated in the cases of copolymers of NIPAM [268] and of poly(NIPAM-*co*-*N*-*tert*-butylacrylamide) (90 : 10, M_w = 160 000 g mol^{-1}) [281]. The hydrophobically modified copolymers formed smaller mesoglobules than homopolymers of similar length and the aggregation number drastically decreased with increasing hydrophobic content [268, 281]. Also, the collapse of a hydrophobically modified macromolecule is expected to proceed faster than that of a homopolymer as a result of the strong attraction between the hydrophobic groups. In some cases the preparation of colloidally stable aggregates of linear homopolymers was not successful, whereas their hydrophobically modified copolymers formed stable mesoglobules upon heating [281]. A model developed by Dobrynin [374] to describe the phase diagram of solutions of associative macromolecules with functional groups/stickers capable of forming physical cross-links may bring further support for the role of hydrophobic groups in mesoglobule stabilisation. An expression derived for a single-chain free energy shows that association between hydrophobes can stabilise a globular state when the energy of pairwise associations is sufficiently large.

5.4 Protein-Like Copolymers

Several block and graft copolymers have been shown to form stable aggregates under thermodynamically poor solvent conditions, as a result of differences in the solubility of different parts of a macromolecule. Whereas in a good solvent the experimentally measured value of A_2 for a copolymer represents the balance of all the multiple interactions, under thermodynamically poor conditions A_2 is mainly determined by the interaction of the groups situated on the polymer–solvent interface. Groups which form the hydrophobic core and are not in a contact with the solvent do not contribute significantly to the solution properties of the copolymer.

The surfaces of mesoglobules of amphiphilic copolymers in water above the LCST are expected to have some hydrophilic character. The hydration layer tightly bound to the chain fragments on the mesoglobule–water interface is believed to trigger short-range repulsion between the hydrophilic layers of two mesoglobules as a result of the specific order of the water molecules near the surface. Repulsion may also arise from the roughness of the mesoglobule surface: thermal motion may induce groups, for example hydrated chain segments, to protrude from the mesoglobule surface, inducing some level of steric repulsion between the particles. Israelachvili et al. [12, 13] have proposed that, in many cases, the observed "hydration force" could be explained by such protruding forces on the molecular level or by undulation forces when the whole surface is in thermal undulating motion, which may also create steric repulsion.

It is generally accepted that in globular proteins the hydrophilic units mainly cover the surface of the globule, thus preventing interprotein association, while hydrophobic units mainly form the core of the globule. Amphiphilic copolymers can mimic the behaviour of biopolymers and, in certain cases, that of globular proteins. Recently, Khokhlov et al. [257–260] proposed that amphiphilic copolymers that remain soluble in an aqueous medium after their transition to the globular state mimic globular proteins and might be called protein-like polymers. The theoretical model of protein-like copolymers ascribes a "memory" to amphiphilic copolymers, stating that a polymer chain tends to reassume the conformation in which it was synthesised, owing to the unique distribution of repeating units along this chain. The synthesis of protein-like copolymers from typical synthetic monomers is difficult, and only a small number of publications have presented data in support of this hypothesis. Examples include PNIPAM copolymers (Sect. 2.2 and Refs. [165, 166, 183, 375]) and PVCL copolymers [174, 175].

6 Conclusions

In this review, we have collected instances under which stable single polymer globules or multipolymeric mesoglobules exhibit remarkable stability against macroscopic phase separation in a thermodynamically poor solvent. Although the focus of the review was on neutral polymers in water, we noted (Sect. 5.1) that the same phenomenon takes place in the case of polymers in organic solvents. The mechanism responsible for this stability against precipitation is still under discussion. Several key factors emerge as important contributors.

The stability of mesoglobular dispersions can be understood in terms of the difference in the characteristic time between collisions of two globules and the time of reptation of a chain on the globule surface. Merging of two globules requires chain reptation, a process estimated to take about as long as the time required for a globule to diffuse over its diameter. Meanwhile, the globules experience an entropic counterforce, which tends to separate them before reptation and entanglement of chains belonging to separate globules can actually take place.

It has been argued that partial vitrification may take place within the core of the globules or mesoglobules. Vitrification drastically decreases the mobility of macromolecules. Collisions of such polymeric particles are purely elastic, precluding merging or coalescence of two globules. In the case of amphiphilic copolymers containing fragments which are strongly attracted to each other core–shell particles may form. Insoluble blocks, segments, or groups form a dense core, restricting chain mobility/reptation and preventing interparticle entanglements. In this scenario the core plays the role of a stabiliser.

The stability of mesoglobules in water may also be ascribed to the presence of a layer of water of hydration bound to the surface of the globules and it is this highly hydrated outer shell that drives the stability of the mesoglobules. The occurrence of such a mechanism is highly likely in the case of mesoglobules of copolymers containing hydrophilic fragments. Whether is it active in the case of homopolymers as well remains to be assessed.

Both experimental evidence and theoretical models are needed to delineate the parameters controlling the properties of "stable globules and mesoglobules" in thermodynamically poor conditions. They may lead to the rational design of families of particles that may find important practical applications in various fields.

References

1. McNaught AD, Wilkinson A (1997) (eds) Compendium of chemical terminology: IUPAC recommendations. The gold book, 2nd edn. Blackwell, Oxford; http://www.iupac.org/publications/compendium/index.html
2. De Gennes PG, Badoz J (1998) Hauraat esineet (fragile objects). Terra cognita, Helsinki
3. Hunter RJ (2001) Foundations of colloid science. Oxford University Press, Oxford
4. Evans DF, Wennerström H (1999) The colloidal domain. Wiley, New York
5. Sonntag H, Strenge K (1987) Coagulation kinetics and structure formation. Plenum, New York
6. Einarson MB, Berg JC (1992) Langmuir 8:2611
7. Sofia SJ, Merrill EW (1997) In: Harris JM, Zaplisky S (eds) Poly(ethylene glycol): chemistry and biological applications. ACS Symposium Series 680. American Chemical Society, Washington, DC, p 342
8. Israelachvili JN, Wennerström H (1990) Langmuir 6:873
9. Derjaguin BV, Churaev NV (1989) Colloids Surf 41:223
10. Besseling NAM (1997) Langmuir 13:2113
11. Marcelja S, Radic N (1976) Chem Phys Lett 42:129
12. Israelachvili JN, Wennerström H (1996) Nature 379:219
13. Leckband D, Israelachvili J (2001) Q Rev Biophys 34:105
14. Del Bene JE, Pople JA (1973) J Chem Phys 58:3605
15. Frank HS (1974) Restrictions for an acceptable model for water structure. In: Luck WAP (ed) Structure of water and aqueous solutions. Verlag Chemie & Physik Verlag, Weinheim, p 9
16. Symons MCR (1975) Philos Trans R Soc Lond Ser B 272:13
17. Ben-Naim A (1974) Water and aqueous solutions: introduction to a molecular theory. Plenum, New York
18. Privalov PL, Gill SJ (1988) Adv Protein Chem 39:191
19. Muller N (1990) Acc Chem Res 23:23
20. Lee B, Graziano G (1996) J Am Chem Soc 22:5163
21. Liu K, Cruzan JD, Saykally RJ (1996) Science 271:929
22. Klug DD (2001) Science 294:2305
23. Stace A (2001) Science 294:1292
24. Zwier TS (2004) Science 304:1119
25. Lower S (2001) http://www.chem1.com/acad/sci/aboutwater.html

26. Chaplin M (2005) http://www.lsbu.ac.uk/water and references therein
27. Klotz IM (1965) Fed Proc 24:Suppl 15:S-24
28. Franks HS, Evans MW (1945) J Chem Phys 13:507
29. Kauzmann W (1959) Adv Protein Chem 14:1
30. Franks F (ed) Water: a comprehensive treatise, vol 4. Plenum, New York
31. Tanford C (1973) The hydrophobic effect: formation of micelles and biological membranes. Wiley, New York
32. Franks F, Eagland D (1985) CRC Crit Rev Biochem 4:165
33. Ben-Naim A (1980) Hydrophobic interactions. Plenum, New York
34. Molyneux P (1985) Water-soluble synthetic polymers: properties and behaviour. CRC, Boca Raton, FL
35. Pollack GL (1991) Science 251:1323
36. Tanford C (1966) Physical chemistry of macromolecules. Wiley, New York
37. Israelachvili JN (1992) Intermolecular and surface forces. Academic, New York
38. Widom B, Bhimalapuram B, Koga K (2003) Phys Chem Phys 5:3085
39. Southall NT, Dill KA, Haymet ADJ (2002) J Phys Chem B 106:521
40. Glass JE (ed) (1986) Water-soluble polymers: beauty with performance. Advances in chemistry 213. American Chemical Society, Washington, DC
41. Silverstein KAT, Haymet ADJ, Dill KA (2000) J Am Chem Soc 122:8037
42. Silverstein KAT, Haymet ADJ, Dill KA (1999) J Chem Phys 111:8000
43. De Los Rios P, Caldarelli G (2000) Phys Rev E 62:8449
44. Caldarelli G, De Los Rios P (2001) J Biol Phys 27:229
45. Moelbert S, De Los Rios P (2003) Macromolecules 36:5845
46. Wohlfarth C (2004) CRC Handbook of thermodynamic data of aqueous polymer solutions. CRC, Boca Raton, FL
47. Nishio I, Sun ST, Swislow G, Tanaka T (1979) Nature 281:208
48. Baranovskaja IA, Klenin SI, Molotkov VA (1982) Vysokomol Soedin Ser B 24:607
49. Winnik FM, Ringsdorf H, Venzmerand J (1990) Macromolecules 23:2415
50. McCormick CL (ed) (2001) Stimuli-responsive water-soluble and amphiphilic polymers. ACS symposium series 780. American Chemical Society, Washington, DC
51. Seok Gil E, Hudson S (2004) Prog Polym Sci 29:1173
52. Flory PJ (1949) J Chem Phys 17:303
53. Flory PJ (1953) Principles of polymer chemistry. Cornell University Press, Ithaca, NY
54. De Gennes PG (1976) Scaling concepts in polymer science. Cornell University Press, Ithaca, New York
55. Grosberg AY, Khokhlov AR (1994) Statistical physics of macromolecules. AIP, New York
56. Sanchez JC (1979) Macromolecules 12:980
57. Yamakawa H (1993) Macromolecules 26:5061
58. Yamakawa H, Abe F, Einaga Y (1994) Macromolecules 27:5704
59. Ptitsyn OB, Eizner YY (1965) Biofizika 10:3
60. Lifshitz IM (1968) Zh Eksp Teor Fiz 55:2408
61. Eizner YY (1969) Vysokomol Soedin Ser A 11:364
62. Lifshitz IM (1969) Sov Phys JETP 28:1280
63. Lifshitz IM, Grosberg AY, Khokhlov AR (1978) Rev Mod Phys 50:683
64. Lifshitz IM, Grosberg AY, Khokhlov AR (1979) Usp Fiz Nauk 127:353
65. Tanaka F (1985) J Chem Phys 82:4707
66. Birshtein TM, Pryamitsyn VA (1987) Vysokomol Soedin Ser A 29:1858
67. Grosberg AY, Nechaev SK, Shakhnovich EI (1988) J Phys 49:2095
68. Birshtein TM, Pryamitsyn VA (1991) Macromolecules 24:1554

69. Birshtein TM, Pryamitsyn VA (1991) Macromolecules 24:3468
70. Grosberg AY, Kuznetsov DV (1992) Macromolecules 25:1970
71. Grosberg AY, Kuznetsov DV (1992) Macromolecules 25:1980
72. Grosberg AY, Kuznetsov DV (1992) Macromolecules 25:1991
73. Grosberg AY, Kuznetsov DV (1992) Macromolecules 25:1996
74. Grosberg AY, Kuznetsov DV (1993) Macromolecules 26:4249
75. Grosberg AY, Khokhlov AR (1997) Giant molecules: here, there, and everywhere... Academic, San Diego
76. Witelski TP, Grosberg AY, Tanaka T (1998) J Chem Phys 108:9144
77. Halperin A, Goldbart P (2000) Phys Rev E 61:565
78. Shmakov SL (2002) Vysokomol Soedin Ser B 44:2223
79. Klushin LI (1998) J Chem Phys 108:7917
80. Yu J, Wang Z, Chu B (1992) Macromolecules 25:1618
81. Chu B, Ying Q (1994) Polym Prepr Am Chem Soc Div Polym Chem 35:181
82. Chu B, Yu J, Wang Z (1993) Prog Colloid Polym Sci 91:142
83. Chu B, Ying Q, Grosberg AY (1995) Macromolecules 28:180
84. Chu B, Wu C (1996) Vysokomol Soedin Ser A 38:574
85. Koningsveld R, Stockmayer WH, Nies E (2001) Polymer phase diagrams, a textbook. Oxford University Press, Oxford
86. Swislow G, Sun ST, Nishio I, Tanaka T (1980) Phys Rev Lett 44:796
87. Sun ST, Nishio I, Swislow G, Tanaka T (1980) J Chem Phys 73:5971
88. Nishio I, Swislow G, Sun ST, Tanaka T (1982) Nature (300:243
89. Miyaki Y, Fujita H (1981) Polym J 13:749
90. Nerger D, Eisele M, Kajiwara K (1983) Polym Bull 10:182
91. Park IH, Wang QW, Chu B (1987) Macromolecules 20:1965
92. Chu B, Xu R, Wang Z, Zuo J (1988) J Appl Crystallogr 21:707
93. Vshivkov SA, Safronov AP (1995) Vysokomol Soedin Ser B 37:1779
94. Vshivkov SA, Safronov AP (1997) Macromol Chem Phys 198:3015
95. Štèpánek P, Koňák C, Sedlácek B (1985) In: Sedlácek B (ed) Physical optics of dynamic phenomena and processes in macromolecular systems. Proceedings of the 27th microsymposium on macromolecules. de Gruyter, Berlin, p 271
96. Picarra S, Duhamel J, Fedorov A, Martinho JMG (2004) J Phys Chem B 108:12009
97. Baysal BM, Karasz FE (2003) Macromol Theory Simul 12:627
98. Yamakawa H (1971) Modern theory of polymer solutions. Harper & Row, New York
99. Williams C, Brochard F, Frich CH (1981) Annu Rev Phys Chem 32:433
100. Fujita H (1990) Polymer solutions. Elsevier, New York
101. Des Cloizeaux JJ, Jannink G (1990) Polymer solutions: their modelling and structure. Clarendon Press, Oxford
102. Klenin VJ (1999) Thermodynamics of systems containing flexible chain polymers. Elsevier, Amsterdam
103. Di Marzio (1999) Prog Polym Sci 24:329
104. Bercea M, Ioan C, Ioan S, Simionescu BC, Simionescu CI (1999) Prog Polym Sci 24:379
105. Keller A, Cheng SZD (1998) Polymer 39:4461
106. Keller A (1995) Macromol Symp 98:1
107. Keller A (1995) Faraday Discuss 101:1
108. Perzynski R, Adam M, Delsanti M (1982) J Phys 43:129
109. Perzynski R, Delsanti M, Adam M (1984) J Phys 45:1765
110. Hadjichristidis N, Bitterlin E, Fetters L, Rosenblum W, Nonidez W, Nan S, Mays J (1994) Polymer 35:4638

111. Fujishige S (1987) Polym J 19:297
112. Fujishige S, Kubota K, Ando I (1989) J Phys Chem 93:3311
113. Kubota K, Fujishige S, Ando I (1990) J Phys Chem 94:5154
114. Ricka J, Meewes M, Nyffenegger R, Binkert T (1990) Phys Rev Lett 65:657
115. Meewes M, Ricka J, de Silva M, Nyffenegger R, Binkert T (1991) Macromolecules 24:5811
116. Tiktopulo EI, Bychkova VE, Ricka J, Ptitsyn OB (1994) Macromolecules 27:2879
117. Tiktopulo EI, Uversky VN, Lushchik VB, Klenin SI, Bychkova VE, Ptitsyn OB (1995) Macromolecules 28:7519
118. Napper DH (1995) Macromol Symp 98:911
119. Wu C, Zhou S (1995) Macromolecules 28:8381
120. Wu C, Zhou S (1995) Macromolecules 28:5388
121. Wang X, Qiu X, Wu C (1998) Macromolecules 31:2972
122. Wu C, Wang X (1998) Phys Rev Lett 80:4092
123. Lerman LS (1971) Proc Natl Acad Sci USA 68:1886
124. Lerman LS, Allen SL (1973) Cold Spring Harbor Symp Quantum Biol 38:59
125. Klenin SI, Baranovskaya IA, Valueva SV (1993) Vysokomol Soedin Ser A 35:838
126. Klenin SI, Baranovskaya IA, Valueva SV (1993) Polym Sci A 35:934
127. Kipper AI, Valueva SV, Bykova EN, Samarova OE, Rumyanceva NV, Klenin SI (1994) Vysokomol Soedin Ser A 36:976
128. Vasilevskaya VV, Khokhlov AR, Matsuzawa Y, Yoshikawa K (1995) J Chem Phys 102:6595
129. Yoshikawa K (1996) Macromol Symp 106:367
130. Vasilevskaya VV, Khokhlov AR, Kidoaki S, Yoshikawa K (1997) Biopolymers 41:51
131. Klenin SI, Baranovskaya IA, Aseyev VO (1996) Macromol Symp 106:205
132. Aseyev VO, Tenhu H, Klenin SI (1998) Macromolecules 31:7717
133. Aseyev VO, Tenhu H, Klenin SI (1999) Macromolecules 32:1838
134. Aseyev VO, Tenhu H, Klenin SI (1999) Polymer 40:1173
135. Aseyev VO, Klenin SI, Tenhu H, Grillo I, Geissler E (2001) Macromolecules 34:3706
136. Serhatli E, Serhatli M, Baysal BM, Karasze FE (2002) Polymer 43:5439
137. Lee MJ, Green MM, Mikeš F, Morawetz H (2002) Macromolecules 35:4216
138. Kiriy A, Gorodyska G, Minko S, Jaeger W, Štěpánek P, Stamm M (2002) J Am Chem Soc 124:13454
139. Vasilevskaya VV, Khalatur PG, Khokhlov AR (2003) Macromolecules 36:10103
140. Dobrynin AV, Rubinstein M, Obukhov SP (1996) Macromolecules 29:2974
141. Chan K, Pelton R, Zhang J (1999) Langmuir 15:4018
142. Pelton R (2000) Adv Colloid Interface Sci 85:1
143. Pelton R (2004) Macromol Symp 207:57
144. Gorelov AV, Du Chesne A, Dawson KA (1997) Physica A 240:443
145. Dawson KA, Gorelov AV, Timoshenko EG, Kuznetsov YA, Du Chesne A (1997) Physica A 244:68
146. Laukkanen A, Valtola L, Winnik FM, Tenhu H (2004) Macromolecules 37:2268
147. Aseyev V, Hietala S, Laukkanen A, Nuopponen M, Confortini O, Du Prez FE, Tenhu H (2005) Polymer 46:7118
148. Solis FJ, Weiss-Malik R, Vernon B (2005) Macromolecules 38:4456
149. Taylor LD, Cerankowski LD (1975) J Polym Sci Part A Polym Chem 13:2551
150. Schild GH (1992) Prog Polym Sci 17:163
151. Kirsh YE (1998) Water soluble poly-*N*-vinylamides: synthesis and physicochemical properties. Wiley, Chichester
152. Kirsh YE (1993) Prog Polym Sci 18:519

153. Horne RA, Almeida JP, Day AF, Yu NT (1971) J Colloid Interface Sci 35:77
154. Nishi T, Kwei K (1975) Polymer 16:285
155. Maeda H (1994) J Polym Sci Polym Phys 32:91
156. Maeda H (1995) Macromolecules 28:5156
157. Meeussen F, Nies E, Berghmans H, Verbrugghe S, Goethals E, Du Prez F (2000) Polymer 41:8597
158. Van Durme K, Verbrugghe S, Du Prez FE, Van Mele B (2004) Macromolecules 37:1054
159. Afroze F, Nies E, Berghmans H (2000) J Mol Struct 554:55
160. Van Durme K, Van Assche G, Van Mele B (2004) Macromolecules 37:9596
161. Schäfer-Soenen H, Moerkerke R, Berghmans H, Koningsveld R, Dušek K, Šolc K (1997) Macromolecules 30:410
162. Moerkerke R, Meeussen F, Koningsveld R, Berghmans H, Mondelaers W, Schacht E, Dušek K, Šolc K (1998) Macromolecules 31:2223
163. Meeussen F, Bauwens Y, Moerkerke R, Nies E, Berghmans H (2000) Polymer 41:3737
164. Swier S, Van Durme K, Van Mele B (2003) J Polym Sci Part B Polym Phys 41:1824
165. Virtanen J, Baron C, Tenhu H (2000) Macromolecules 33:336
166. Virtanen J, Tenhu H (2000) Macromolecules 33:5970
167. Virtanen J, Lemmetyinen H, Tenhu H (2001) Polymer 42:9487
168. Virtanen J, Tenhu H (2001) J Polym Sci Part A Polym Chem 39:3716
169. Virtanen J, Holappa S, Lemmetyinen H, Tenhu H (2002) Macromolecules 35:4763
170. Virtanen J (2002) Self-assembling of thermally responsive block and graft copolymers in aqueous solutions. Academic dissertation. Yliopistopaino, University of Helsinki. http://ethesis.helsinki.fi/julkaisut/mat/kemia/vk/virtanen
171. Zhang W, Shi L, Wu K, An Y (2005) Macromolecules 38:5743
172. Motokawa R, Morishita K, Koizumi S, Nakahira T, Annaka M (2005) Macromolecules 38:5748
173. Deng Y, Pelton R (1995) Macromolecules 28:4617
174. Lozinsky VI, Simenel IA, Kurskaya EA, Kulakova VK, Grinberg BY, Lubovik AS, Galaev IY, Mattiasson B, Khokhlov AR (2000) Dokl Akad Nauk 375:637
175. Lozinsky VI, Simenel IA, Kulakova VK, Kurskaya EA, Babushkina TA, Klimova TP, Burova TV, Dubovik AS, Grinberg VY, Galaev IY, Mattiasson B, Khokhlov AR (2003) Macromolecules 36:7308
176. Qiu X, Kwan CMS, Wu C (1997) Macromolecules 30:6090
177. Laukkanen A, Hietala S, Maunu SL, Tenhu H (2000) Macromolecules 33:8703
178. Laukkanen A, Wiedmer SK, Varjo S, Riekkola ML, Tenhu H (2002) Colloid Polym Sci 280:65
179. Laukkanen A, Winnik FM, Tenhu H (2005) Macromolecules 38:2439
180. Laukkanen A, Valtola L, Winnik FM, Tenhu H (2005) Polymer 46:7055
181. Laukkanen A (2005) Thermally responsive polymers based on *N*-vinylcaprolactam and an amphiphilic macromonomer. Academic Dissertation. Yliopistopaino, University of Helsinki. http://ethesis.helsinki.fi/julkaisut/mat/kemia/vk/laukkanen
182. Verbrugghe S, Laukkanen A, Aseyev V, Tenhu H, Winnik FM, Du Prez FE (2003) Polymer 44:6807
183. Siu MH, Cheng HC, Wu C (2003) Macromolecules 36:6588
184. Shtanko NI, Kabanov VY, Apel PY, Yoshida M, Vilenskii AI (2000) J Membr Sci 179:155
185. Nuopponen M, Ojala J, Tenhu H (2004) Polymer 45:3643
186. American Cyanamid Company (1963) *N*-Isopropylacrylamide. Brochure IC3-1354-500-4/63
187. Scarpa JS, Muller DD, Klotz IM (1967) J Am Chem Soc 89:6024

188. Heskins M, Guillet JE (1969) J Macromol Sci Chem 2:1441
189. Graziano G (2000) Int J Biol Macromol 27:89
190. Zhang G, Wu C (2001) Phys Rev Lett 86:822
191. Liang L, Rieke PC, Fryxell GE, Liu J, Engehard MH, Alford KL (2000) J Phys Chem B 104:11667
192. Zhang W, Zou S, Wang C, Zhang X (2000) J Phys Chem B 104:10258
193. Wang X, Wu C (1999) Macromolecules 32:4299
194. Zhang G, Wu C (2001) J Am Chem Soc 123:1376
195. Hu T, Wu C (2001) Macromolecules 34:6802
196. Greenwood R, Kendall K, Ritchie S, Snowden MJ (2000) J Eur Ceram Soc 20:1707
197. Hatto N, Cosgrove T, Snowden MJ (2000) Polymer 41:7133
198. Lowe TL, Virtanen J, Tenhu H (1999) Langmuir 15:4259
199. Wolf BA (1985) Pure Appl Chem 57:323
200. Kujawa P, Watanabe H, Tanaka F, Winnik FM (2005) Eur Phys J E 17:129
201. Otake K, Inomata H, Konno M, Saito S (1990) Macromolecules 23:283
202. Inomata H, Goto S, Saito S (1990) Macromolecules 23:4887
203. Bae YH, Okano T, Kim SW (1990) J Polym Sci Part B Polym Phys 28:923
204. Feil H, Bae YH, Feijen J, Kim SW (1993) Macromolecules 26:2496
205. Shibayama M, Mizutani S, Nomura S (1996) Macromolecules 29:2019
206. Suetoh Y, Shibayama M (2000) Polymer 41:505
207. Tamai Y, Tanaka H, Nakanishi K (1996) Macromolecules 29:6750
208. Ebara M, Yamato M, Motohiro H, Aoyagi T, Kikuchi A, Sakai K, Okano T (2003) Biomacromolecules 4:344
209. Kujawa P, Winnik F (2001) Macromolecules 34:4130
210. Lin LN, Brandts JF, Brandts M, Plotnikov V (2002) Anal Biochem 302:144
211. Cho EC, Lee J, Cho K (2003) Macromolecules 36:9929
212. Reading M (1993) Trends Polym Sci 1:248
213. Reading M, Luget A, Wilson R (1993) Thermochim Acta 238:295
214. Burchard W (1998) Adv Polym Sci 143:173
215. Timoshenko EG, Kuznetsov YA (2000) Condens Matter arXiv:cond-mat/0011386
216. Timoshenko EG, Kuznetsov YA (2000) J Chem Phys 112:8163
217. Timoshenko EG, Kuznetsov YA (2000) Prog Colloid Polym Sci 115:117
218. Gao J, Frisken BJ (2003) Langmuir 19:5212
219. Gao J, Frisken BJ (2003) Langmuir 19:5217
220. Gao J, Frisken BJ (2005) Langmuir 21:545
221. Thunemann AF, Beyermann J, Kukula H (2000) Macromolecules 33:5906
222. Gohy JF, Varshney SK, Jerome R (2001) Macromolecules 34:3361
223. Sedlak M, Antonietti M, Cölfen H (1998) Macromol Chem Phys 199:247
224. Qi LM, Cölfen H, Antonietti M (2000) Angew Chem Int Ed Engl 39:604
225. Martin TJ, Prochazka K, Munk P, Webber SE (1996) Macromolecules 29:6071
226. Bütün V, Billingham NC, Armes SP (1998) J Am Chem Soc 120:11818
227. Zhu PW (2004) J Mater Sci Mater Med 15:567
228. Mathur A, Drescher B, Scranton A, Klier J (1998) Nature 392:367
229. Bütün V, Billingham N, Armes S (1997) Chem Commum 7:671
230. Harada A, Kataoka K (1999) J Am Chem Soc 121:9241
231. Dash P, Toncheva V, Schacht E, Seymour L (1997) J Controlled Release 48:269
232. Bronich T, Nehls A, Eisenberg A, Kabanov V, Kabanov A (1999) Colloids Surf B 16:243
233. Nishiyama N, Yokoyama M, Aoyagi T, Okano T, Sakurai Y, Kataoka K (1999) Langmuir 15:377
234. Kataoka K, Ishihara A, Harada A, Miyazaki H (1998) Macromolecules 31:6071

235. Kataoka K, Harada A, Nagasaki Y (2001) Adv Drug Delivery Rev 47:113
236. Holappa S, Andersson T, Kantonen L, Plattner P, Tenhu H (2003) Polymer 44:7907
237. Holappa S, Kantonen L, Winnik F, Tenhu H (2004) Macromolecules 37:7008
238. Andersson T, Holappa S, Aseyev V, Tenhu H (2003) J Polym Sci Part A Polym Chem 41:1904
239. Andersson T, Aseyev V, Tenhu H (2004) Biomacromolecules 5:1853
240. Kabanov V, Zezin A (1984) Pure Appl Chem 56:343
241. Bakeev K, Izumrudov V, Kuchanov S, Zezin A, Kabanov V (1992) Macromolecules 25:4249
242. Dautzenberg H (2001) In: Radeva T (ed) Physical chemistry of polyelectrolytes. Dekker, New York, p 743
243. Zintchenko A, Rother G, Dautzenberg H (2003) Langmuir 19:2507
244. Kawaguchi S, Winnik MA, Ito K (1996) Macromolecules 29:4465
245. Alexander S (1977) J Phys 38:983
246. De Gennes PG (1980) Macromolecules 13:1069
247. Qiu X, Wu C (1997) Macromolecules 30:7921
248. Topp MDC, Leunen IH, Dijkstra PJ, Tauer K, Schellenberg C, Feijen J (2000) Macromolecules 33:4986
249. Topp MDC, Dijkstra PJ, Talsma H, Feijen J (1997) Macromolecules 30:8518
250. Zhu PW, Napper DH (1999) Macromolecules 32:2068
251. Liang D, Zhou S, Song L, Zaitsev VS, Chu B (1999) Macromolecules 32:6326
252. Berlinova IV, Dimitrov IV, Vladimirov NG, Samichkov V, Ivanov Y (2001) Polymer 42:5963
253. Lin HH, Cheng YL (2001) Macromolecules 34:3710
254. Bergbreiter DE, Case BL, Liu YS, Caraway JW (1998) Macromolecules 31:6053
255. Soundararajan S, Reddy BSR (1991) J Appl Polym Sci 43:251
256. Nguyen AL, Luong JHT (1989) Biotechnol Bioeng 34:1186
257. Khokhlov AR, Khalatur PG (1998) Physica A 249:253
258. Khalatur PG, Ivanov VI, Shusharina NP, Khokhlov AR (1998) Russ Chem Bull 47:855
259. Khokhlov AR, Ivanov VA, Shusharina NP, Khalatur PG (1998) Engineering of synthetic copolymers: protein-like copolymers. In: Yonezawa F, Tsuji K, Kaij K, Doi M, Fujiwara T (eds) The physics of complex liquids. World Scientific, Singapore, p 155
260. Khokhlov AR, Khalatur PG (2004) Curr Opin Solid State Mater Sci 8:3
261. Zhang G, Winnik F, Wu C (2003) Phys Rev Lett 90:035506
262. Zhu P, Napper DH (1994) J Colloid Interface Sci 164:489
263. Zhu P, Napper DH (1994) J Colloid Interface Sci 168:380
264. Chen MQ, Serizawa T, Li M, Akashi M (2003) Polym J 35:901
265. Hellweg T, Dewhurst CD, Eimer W, Kratz K (2004) Langmuir 20:4330
266. Zhang W, Zhou X, Li H, Fang Y, Zhang G (2005) Macromolecules 38:909
267. Picarra S, Martinho JMG (2001) Macromolecules 34:53
268. Wu C, Li W, Zhu XX (2004) Macromolecules 37:4989
269. Wu C (2003) Chin J Polym Sci 21:117
270. Siu MH, Liu HY, Zhu XX, Wu C (2003) Macromolecules 36:2103
271. Solomon OF, Corciovei M, Ciuta I, Boghina C (1968) J Appl Polym Sci 12:1835
272. Vihola H, Laukkanen A, Valtola L, Tenhu H, Hirvonen J (2005) Biomaterials 26:3055
273. Galaev IY, Mattiasson B (1993) Enzyme Microb Technol 15:354
274. Lebedev V, Török G, Cser L, Treimer W, Orlova D, Sibilev A (2003) J Appl Crystallogr 36:967
275. Lebedev VT, Török G, Cser L, Kali G, Kirsh YE, Sibilev AI, Orlova DN (2001) Physica B 297:50

276. Lau ACW, Wu C (1999) Macromolecules 32:581
277. Makhaeva EE, Tenhu H, Khokhlov AR (1998) Macromolecules 31:6112
278. Maeda Y, Nakamura T, Ikeda I (2002) Macromolecules 35:217
279. Qiu Q, Somasundaran P, Pethica BA (2002) Langmuir 18:3482
280. Anufrieva EV, Krakovyak MG, Ananieva TD, Lushchik VB, Nekrasova TN, Papukova KP, Sheveleva TV (2002) Vysokomol Soedin Ser A 44:1530
281. Anufrieva EV, Ananieva TD, Krakovyak MG, Lushchik VB, Nekrasova TN, Smyslov RJ, Sheveleva TV (2005) Vysokomol Soedin Ser A 47:1
282. Freedman HH, Mason JP, Medalia AI (1958) J Org Chem 23:76
283. Gyuot A (2002) Macromol Symp 179:105
284. Guyot A, Tauer K, Asua JM, Van Es S, Gauthier C, Hellgren AC, Sherrington DC, Montoya-Goni A, Sjoberg M, Sindt O, Vidal F, Unzue M, Schoonbrood H, Shipper E, Lacroix-Desmazes P (1999) Acta Polym 50:57
285. Paleos CM (ed) (1992) Polymerization in organized media. Gordon and Breach, New York
286. Capek I (2000) Adv Colloid Interface Sci 88:295
287. Capek I (1999) Adv Polym Sci 145:1
288. Asua JM, Schoonbrood HAS (1998) Acta Polym 49:671
289. Ito K, Kawaguchi S (1998) Adv Polym Sci 142:129
290. Velichkova RS, Christova DC (1995) Prog Polym Sci 20:819
291. Ito K (1998) Prog Polym Sci 23:581
292. Varadaraj R, Branham KD, McCormick CL, Bock J (1994) Analysis of hydrophobically associating copolymers utilizing spectroscopic probes and labels. In: Dubin P, Bock J, Davis R, Schulz DN, Thies C (eds) Macromolecular complexes in chemistry and biology. Springer-Verlag, Berlin, p 15
293. Taylor KC, Nasr-El-Din HA (1998) J Pet Sci Eng 19:265
294. Schulz DN, Glass JE (eds) (1991) Polymers as rheology modifiers. ACS symposium series 462. American Chemical Society, Washington, DC
295. Kästner U (2001) Colloids Surf A 183:805
296. Van den Hull HJ, Vanderhoff JW (1970) In: Flitch RM (ed) Polymer colloids: proceedings of an American Chemical Society symposium on polymer colloids held in Chicago, IL, 13–18 September. Plenum, New York, p 3 and references therein
297. Daly E, Saunders BR (2000) Langmuir 13:5546
298. Duracher D, Elaissari A, Pichot C (1999) Colloid Polym Sci 277:905
299. Napper DH (1970) J Colloid Interface Sci 33:384
300. Boucher EA, Hines PM (1976) J Polym Sci Polym Phys Ed 14:2241
301. Ottewill RH, Satgurunathan R (1988) Colloid Polym Sci 266:547
302. Ottewill RH, Satgurunathan R (1995) Colloid Polym Sci 273:379
303. Wu X, Pelton RH, Hamielec AE, Woods DR, McPhee W (1994) Colloid Polym Sci 272:467
304. Senff H, Richtering W (2000) Colloid Polym Sci 278:830
305. Bo G, Wesslén B, Wesslén KB (1992) J Polym Sci Part A Polym Chem 30:1799
306. Jannasch P, Wesslén B (1993) J Polym Sci Part A Polym Chem 31:1519
307. Beezer AE, Loh W, Mitchell JC, Royall PG, Smith DO, Tute MS, Armstrong JK, Chowdhry BZ, Leharne SA, Eagland D, Crowtherg NJ (1994) Langmuir 10:4001
308. Kujawa P, Goh CCE, Calvet D, Winnik F (2001) Macromolecules 34:6387
309. Kirsh YE, Yanul NA, Kalninsh KK (1999) Eur Polym J 35:305
310. Ben-Naim A (1980) Hydrophobic interactions. Plenum, New York
311. Ringsdorf H, Venzmer J, Winnik FM (1991) Macromolecules 24:1678
312. Francis MF, Piredda M, Winnik FM (2003) J Controlled Release 93:59

313. Lakowicz JR (1999) Principles of fluorescence spectroscopy, 2nd edn. Kluwer/Plenum, New York
314. Berlman IB (1973) Energy transfer parameters of aromatic compounds. Academic, New York
315. Winnik FM (1990) Polymer 31:2125
316. Gao Y, Au-Yeung SCF, Wu C (1999) Macromolecules 32:3674
317. Kuz'kina EF, Pashkin II, Markvicheva EA, Kirsh YE, Bakeeva IV, Zubov VP (1996) Khim-Farm Zh 1:39
318. Markvicheva EA, Kuz'kina EF, Pashkin II, Plechko TN, Kirsh YE, Zubov VP (1991) Biotechnol Tech 5:223
319. Maeda H (1994) J Polym Sci Part B Polym Phys 32:4299
320. Maeda Y (2001) Langmuir 17:1737
321. Spěváček J, Hanyková L, Ilavský M (2001) Macromol Symp 166:231
322. Hanyková L, Spěváček J, Ilavský M (2001) Polymer 42:8607
323. Spěváček J, Hanyková L (2003) Macromol Symp 203:229
324. Spěváček J, Hanyková L, Starovoytova L (2004) Macromolecules 37:7710
325. Yang Y, Zeng F, Xie X, Tong Z, Liu X (2001) Polym J 33:399
326. Suzuki M, Hirasa O (1993) Adv Polym Sci 110:241
327. De Rossi D, Kajiwara K, Osada Y, Yamauchi A (eds) (1991) Polymer gels: fundamental and biomedical applications. Plenum, New York
328. Ichijo H, Kishi R, Hirasa O, Takiguchi Y (1994) Polym Gels Networks 2:315
329. Arndt KF, Schmidt T, Menge H (2001) Macromol Symp 164:313
330. Arndt KF, Schmidt T, Reichelt R (2001) Polymer 42:6785
331. Nies E, Ramzi A, Berghmans H, Li T, Heenan RK, King SM (2005) Macromolecules 38:915
332. Burchard W (1977) Macromolecules 18:919
333. Smoluchowski M (1917) Z Phys Chem 92:129
334. Chuang J, Grosberg AY, Tanaka T (2000) J Chem Phys 112:6434
335. Anufrieva EV, Volkenstein MV, Gotlib YY, Krakovyak MG, Pautov VD, Stepanov VV, Skorokhodov SS (1972) Dokl Akad Nauk SSSR 207:1379
336. Anufrieva EV, Krakovyak MG, Gromova RA, Lushchik VB, Ananieva TD, Sheveleva TV (1991) Dokl Akad Nauk SSSR 319:895
337. Marchetti M, Prager S, Cussler EL (1990) Macromolecules 23:3445
338. Lau KF, Dill KA (1989) Macromolecules 22:3986
339. Baulin VA, Zhulina EB, Halperin A (2003) J Chem Phys 119:10977
340. Kauzmann W (1948) Chem Rev 43:219
341. Gibbs JH, DiMarzio EA (1958) J Chem Phys 28:373
342. Hikmet RM, Callister S, Keller A (1988) Polymer 29:1378
343. Callister S, Keller A, Hikmet RM (1990) Makromol Chem Macromol Symp 39:19
344. Koningsveld R, Stockmayer WH, Nies E (1990) Makromol Chem Macromol Symp 39:1
345. Arnauts J, Berghmans H (1987) Polymer Commun 28:66
346. Vandeweerdt P, Berghmans H, Tervoort Y (1991) Macromolecules 24:3547
347. Arnauts J, Berghmans H, Koningsveld R (1993) Makromol Chem 194:77
348. Aerts L, Kunz M, Berghmans H, Koningsveld R (1993) Makromol Chem 194:2697
349. Berghmans H, Deberdt F (1994) Philos Trans R Soc Lond A 348:117
350. Berghmans S, Mewis J, Berghmans H, Meijer H (1995) Polymer 36:3085
351. Berghmans S, Berghmans H, Mewis J, Meijer HEH (1996) J Membr Sci 116:171
352. De Cooman R, Vandeweerdt P, Berghmans H, Koningsveld R (1996) J Appl Polym Sci 60:1127

353. De Rudder, Berghmans H, Arnauts J (1999) Polymer 40:5919
354. Khalatur PG, Khokhlov AR, Vasilevskaya VV (1994) Macromol Theory Simul 3:939
355. Milchev A, Binder K (1994) Europhys Lett 26:671
356. Zhou Y, Hall CK, Karplus M (1996) Phys Rev Lett 77:2822
357. Zhou Y, Karplus M, Wichert JM, Hall CK (1997) J Chem Phys 107:10691
358. Paul W, Müller M (2001) J Chem Phys 115:630
359. Wittkop M, Kreitmeier S, Göritz D (1996) J Chem Phys 104:3373
360. Kreitmeier S, Wittkop M, Göritz D (1999) Phys Rev E 59:1982
361. Kreitmeier S (2000) J Chem Phys 112:6925
362. Du R, Grosberg AY, Tanaka T, Rubinstein M (2000) Phys Rev Lett 84:2417
363. Rostiashvili VG, Migliorini G, Vigis TA (2001) Phys Rev E 64:051112
364. Garel T, Orland H (1988) Europhys Lett 6:597
365. Shakhnovich EI, Gutin AM (1989) Biophys Chem 34:187
366. Sfatos CD, Gutin AM, Shakhnovich EI (1993) Phys Rev E 48:465
367. Pande VS, Grosberg AY, Tanaka T (2000) Rev Mod Phys 72:259
368. Geissler PL, Shakhnovich EI, Grosberg AY (2004) Phys Rev E 70:021802
369. Salmerón Sánchez M, Hanyková L, Ilavský M, Monleón Pradas M (2004) Polymer 45:4087
370. Starovoytova L, Spěváček J, Hanyková L, Ilavský M (2003) Macromol Symp 203:239
371. Wu C, Zhou SQ (1996) Phys Rev Lett 77:3053
372. Tanaka H (1993) Phys Rev Lett 71:3158
373. Tanaka H (1992) Macromolecules 25:6377
374. Dobrynin AV (2004) Macromolecules 37:3881
375. Siu MH, Zhang G, Wu C (2002) Macromolecules 35:2723

[illegible] H, Arends [illegible] (1995) Polymer 36:5919

[illegible] (1994) Macromol Theory Simul [illegible]

[illegible] (1994) Europhys Lett [illegible]

[illegible] Phys Rev Lett 72:2422

[illegible] J Chem Phys 107 [illegible]

[illegible]

Adv Polym Sci (2006) 196: 87–127
DOI 10.1007/12_053

Published online: 10 November 2005

Approaches to Chemical Synthesis of Protein-Like Copolymers

Vladimir I. Lozinsky

Institute of Organoelement Compounds, Russian Academy of Sciences, Vavilov St. 28, 119991 Moscow, Russia
loz@ineos.ac.ru

Abstract Protein-like copolymers were first predicted by computer-aided biomimetic design. These copolymers consist of comonomer units of differing hydrophilicity/hydrophobicity. Heterogeneous blockiness, inherent in such copolymers, promotes chain folding with the formation of specific spatial packing: a dense core consisting of hydrophobic units and a polar shell formed by hydrophilic units. This review discusses the approaches, those that have already been described and potential approaches to the chemical synthesis of protein-like copolymers. These approaches are based on the use of macromolecular precursors as well as the appropriate monomers. In addition, some specific physicochemical properties of protein like copolymers, especially their solution behaviour in aqueous media, are considered.

Keywords Chemical synthesis · Heterogeneous blockiness · Physicochemical studies · Protein-like copolymers · Solution behaviour

Abbreviations

APS	ammonium persulfate
Cu^{2+}-IDA-sepharose	iminodiacetate-sepharose loaded with Cu^{2+}-ions
DMSO	dimethylsulfoxide
EDTA	ethylenediaminetetracetic acid, disodium salt

GMA	glycidyl methacrylate
HS-DSC	high-sensitivity differential scanning calorimetry
IMCC	immobilized metal chelate chromatography
LCST	lower critical solution temperature
NMDEA	*N*-methyldiethanolamine
NVCl	*N*-vinylcaprolactam
NiPAAm	*N*-isopropylacrylamide
NVIAz	*N*-vinylimidazole
PAAm	poly(acrylamide)
PEO	poly(ethylene oxide)
PPS	potassium persulphate
PST	phase separation threshold
SEC	size-exclusion chromatography

1 Introduction

The concept of conformation-dependent computer design of the protein-like copolymers was first formulated in the works by Khokhlov, Khalatur et al. [1–8]. This concept is based on the so-called biomimetic approach to computing the sequence of monomeric units in a chain, and also evaluating the conformation adopted in aqueous media by the macromolecules consisting of the amphiphilic comonomers with differing hydrophilicity/hydrophobicity. The theoretical studies showed that the characteristic feature of such copolymers is their heterogeneous blockiness (regarding the length of the blocks and their position along the chain), which promotes the specific "protein-like" folding of the chains resulting in the formation of spatial structure resembling the tertiary structure of globular proteins. This structure includes dense core, predominantly consisting of the units with relatively higher hydrophobicity (***H***), and the hydrophilic shell enriched with well-hydrated polar (***P***) units. Owing to the peculiar sequence of ***H*** and ***P*** units, i.e. owing to their specific distribution along the chain, the macromolecules of such copolymers in aqueous solutions should adopt the conformation that will impart to these copolymers certain properties differing from the properties of both regular alternating block-copolymers and random statistical copolymers. In particular, if the blocks containing more hydrophobic ***H***-units are capable of thermally inducing the coil-globule transition, the protein-like copolymer should not precipitate from aqueous solution at temperatures higher than LCST of the corresponding ***H***-homopolymer. This is due to the fact that the hydrophilic shell of the globules prevents intermolecular hydrophobic interactions, thus averting the pronounced association of macromolecules and the formation of coagulation contacts.

After the publication of the first theoretical works related to the problem of protein-like copolymers, the first attempts to verify experimentally these theoretical concepts were undertaken. Several research groups have tried to synthesize such copolymers and to study their behaviour in aqueous solutions. These and subsequent "chemical" works will be reviewed in this article. Here, we should also stress that this review is devoted to the discussion of possible approaches to the preparation of protein-like *HP*-copolymers as real high molecular weight compounds. In addition, just those physicochemical properties of such copolymers will be discussed that allow attribution of the respective copolymers to the protein-like macromolecules.

2 Possible Methods for Synthesis of Protein-Like *HP*-Copolymers

2.1 General Remarks

On the basis of general considerations, it is quite evident that the chemical synthesis of protein-like *HP*-copolymers can be performed by using one of the following three different methods:

i through the chemical modification of pendant reactive groups in appropriate polymeric precursors (the so-called "chemical colouring" approach, which is similar to the chemical imitation of the theoretical approach used in the computer design of protein-like macromolecules [1]);

ii through copolymerization or co-polycondensation) of appropriate monomeric precursors;

iii by the step-by-step synthesis of the pre-assigned primary sequence of monomer units in a chain likewise the ribosomal biosynthesis of proteins [9] or the procedures used in solid-phase polypeptide (polynucleotide) synthesis [10, 11].

The first method (i) can be accomplished as follows:

- either by the partial hydrophilization of the outer regions of the relatively hydrophobic (but soluble) macromolecules with compulsory retention of the compact "folded" conformation of the inner regions (core) of surface-modified (chemically coloured) polymer particles;
- or, vice versa, by the partial hydrophobization of the hydrophilic coils, which can cause the intramolecular conformational inversion giving rise to the location of hydrophobically-modified regions in the interior of polymer particles, and residual hydrophilic regions forming the outer shell of the resulting globules.

Certainly, the latter variants imply the use of aqueous reaction media, and for the case of organosoluble polymers one should use the reaction solvent with better or poorer thermodynamic quality in respect to the initial non-modified and final modified chains, respectively. Particular examples of this method will be described below in Sect. 2.2.

The second (ii) synthetic method is based on the following ideas:

If the polymerization of a nonequimolar mixture of amphiphilic comonomers (with a compulsory excess of more hydrophobic *H*-monomer) in an aqueous solution is initiated at a temperature higher than the phase separation threshold (**PST**) of the reaction system (when it would already contain polymeric products), one can suppose a certain amount of homooligomeric chains would be quickly formed (consisting of the more hydrophobic *H*-units); with the length of the chains being sufficient for the realization of LCST-behaviour. In such a case, this growing, but yet short, thermoresponsive chain should already have collapsed in the early stages of the polymerization process (when the concentration of such chains is still relatively low) with the formation of a small dense globule performing as a "germ" for the creation of a hydrophobic core of a protein-like macromolecule. For instance, it is known that for the well-studied thermoresponsive polymer, poly(*N*-isopropylacrylamide) (**PNiPAAm**), the corresponding critical length does not exceed 10–15 monomer units [12]. Therefore, subsequent polymerization will proceed under the heterogeneous conditions.

A further sequence of comonomer units in the chains to be formed will depend on the reactivity ratio of the comonomers (ratio of r_1 and r_2 parameters) and on the current position of the growing macroradical end. This growing end can be located either inside the core, or on the periphery of the globule. In the more hydrophobic core, the molecules of the more hydrophobic monomer should preferably be attached, thus resulting in the enrichment of such segments with hydrophobic *H*-units. Alternatively, on the periphery of a globule in an aqueous environment the molecules of the more hydrophilic monomer should preferably react, thus giving rise to the enrichment of these segments with hydrophilic *P*-units. Therefore, the formation of protein-like sequences of these different monomeric units in the macromolecular chains is possible, at least, for part of the resulting chains. The theoretical evaluation of this "mechanism" confirmed, in general, its feasibility [8, 13], and in Fig. 1 a typical computer snapshot of the structure of resulting protein-like macromolecules is shown.

The temperature-dependent behaviour of the protein-like macromolecules in the aqueous medium should differ significantly from the solution behaviour of random temperature-sensitive copolymers, as it follows from the theoretical predictions [1]. Aqueous solutions of protein-like copolymers should exhibit higher phase transition temperatures, whereas conformational transition of such macromolecules from coil to globule should not lead to the loss of water-solubility. However, the growth of some copolymer chains might not follow the "favourable" scenario described above. These macromolecules will then possess a more random distribution of the comonomer units. Therefore, their solution behaviour should not differ significantly from that of the "traditional" temperature-sensitive copolymers, provided the amount of hydrophilic units incorporated into the chains is not high enough to in-

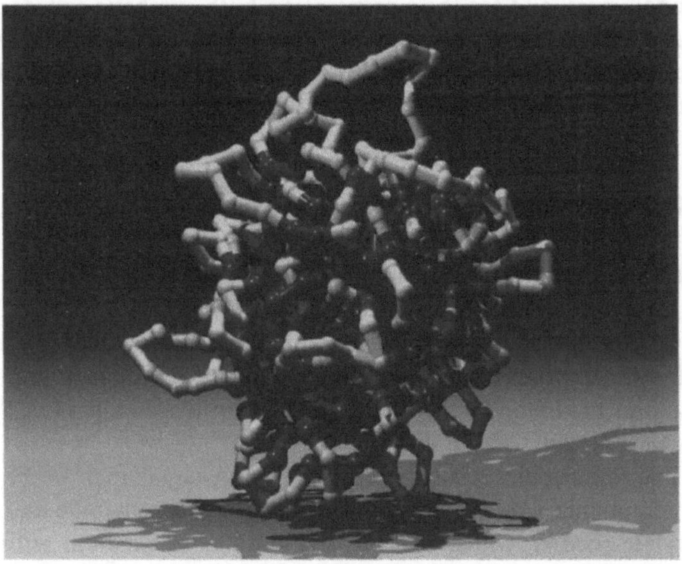

Fig. 1 Typical snapshot of a protein-like copolymer macromolecule, in which ***H*** monomeric units are shown as *dark grey spheres* and ***P*** units are shown as *light grey spheres* (adopted from [13])

hibit the heating-induced phase separation. Similar to approach (i), i.e., the preparation of protein-like copolymers through the chemical modification of polymeric precursors, the second (ii) synthetic approach was also realized experimentally and will be discussed in Sect. 2.3.

The third approach (iii), the solid-phase synthesis, has not been yet implemented for the preparation of protein-like *HP*-copolymers with heterogeneous blockiness, although the possibility—inherent in this method—to form the macromolecular chains with a well-defined chemical sequence of monomeric units is of great interest for the problem. That is why, these possibilities will also be discussed briefly in Sect. 2.4.

2.2 Synthesis of Protein-Like Copolymers Using the Chemical Colouring Approach

The copolymer-based systems possessing the core-shell structure in solutions are known and studied rather well (see, e.g., [14–16]). These copolymers in aqueous media tend to form polymeric micelles, which are often considered as promising drug delivery nano-vehicles [17, 18], i.e., these macromolecular systems are not only of scientific, but also of considerable applied significance. Among such systems there are interesting examples, whose properties are very similar to the properties that should be inherent in the protein-like copolymers. All of these macromolecules possess the primary structure of block-copolymers, and for their preparation both synthetic approaches, i.e., chemical modification of high polymer precursors, and copolymerization of the corresponding monomeric precursors have been implemented. This section of the review deals with the former approach.

2.2.1
Surface Hydrophilization of Folded Macromolecules

In the researches performed by Tenhu and co-authors [19–21], it was shown that by means of chemical modification of the units with reactive pendant groups in thermoresponsive NiPAAm-containing copolymers one can prepare polymeric products, whose solution behaviour resemble in some aspects the behaviour predicted by the theory for protein-like copolymers. Scheme 1 presents the reaction used in these syntheses.

The macromolecular precursor of the graft-copolymer to be created was the copolymer ($M_w = 1.8 \times 10^5$) of *N*-isopropylacrylamide and glycidyl methacrylate (**GMA**), which reacted in the aqueous medium with amino-terminated poly(ethylene oxide) (amino-PEO, $M_w = 6 \times 10^3$) at two temperatures: 15 and 29 °C. Since the temperature of the onset of conformation transition of the "parent" poly(NiPAAm-*co*-GMA) was 28.4 °C, in the first temperature regime the macromolecules of the precursor in the aqueous solution were in a coil conformation, and grafting took place virtually randomly along the chain. On the other hand, at 29 °C (the second reaction temperature) the precursor chains acquired the more compact globular conformation, therefore the modification of epoxide groups was supposed to occur preferably on a periphery of the globules, thus giving rise to the formation of a hydrophilic shell of grafted PEO chains wrapping the collapsed hydrophobic globular core.

Indeed, it turned out that although the average amount of PEO lateral tails per one macromolecule of poly[(NiPAAm-*co*-GMA)-*g*-PEO] prepared at 15 and 29 °C differed insignificantly (6 and 7 grafts, respectively), the temperature-dependent solution behaviour of these two copolymers was found to exhibit some differences. For instance, the onset temperature of the DSC peak for the latter copolymer was somewhat higher, and the rise in turbidity of its aqueous solution upon heating commenced at a temperature about 1.5 °C higher [19]. This testified to certain protective effects of the grafted PEO tails against the heating-induced aggregation of core-shell type macromolecules. However, the absolute magnitudes of the differences in temperatures observed on respective DSC traces and on turbidity curves were not large ("... differences in the critical temperatures were small, but clearly observable and reproducible ..." [20]), thus pointing to not so dramatic distinctions in the properties of these two kinds of graft-copolymers formed upon the modification of macromolecular precursor either in a coil, or in a globular conformation. Besides, light scattering studies showed the smaller

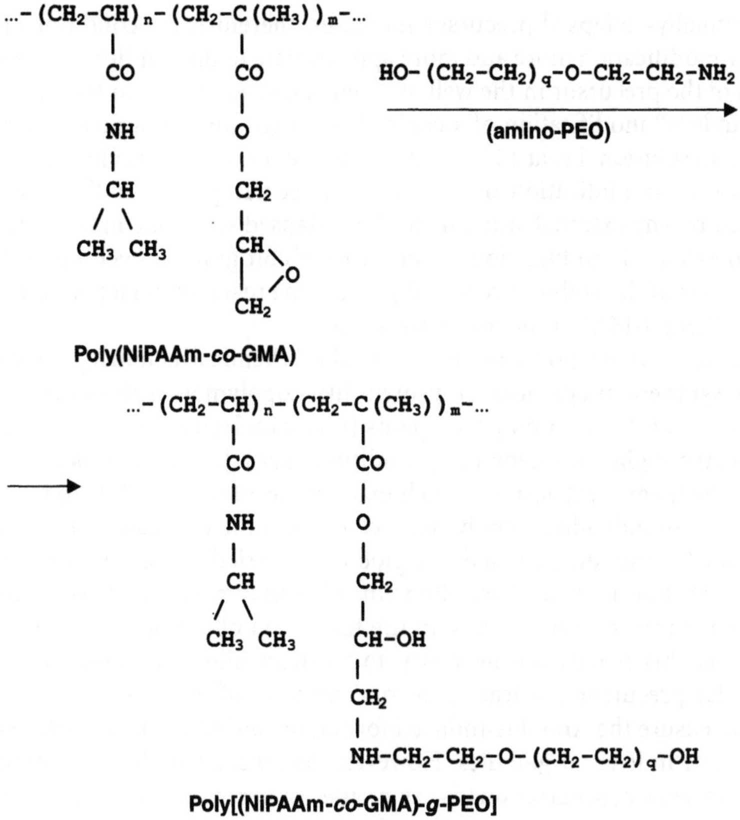

Scheme 1

size of the thermally-collapsed aggregates of graft-copolymer prepared at 29 °C compared to the size of the aggregates formed by the copolymer prepared at 15 °C, when the polymer solutions were examined at 45 °C [20]. This meant the higher density of the aggregates formed at a given elevated temperature from poly[(NiPAAm-*co*-GMA)-*g*-PEO] produced at 29 °C occurred because of a denser hydrophobic core than in the case of aggregates originating from the copolymer prepared in solution at 15 °C, i.e., below the critical temperature. However, again, the differences were not considerable.

It is thought that one of the main reasons for such not very pronounced differences in the properties of the graft-copolymers under discussion could be the process of a gradual (in the course of the grafting process) increase in hydrophilicity of regions of collapsed globules neighbouring the grafts as PEO fragments were attached to the poly(NiPAAm-*co*-GMA) backbone. Such a local surface hydrophilization should result in an additional swelling of modified regions and, as a consequence, in the enhancement of penetration of amino-PEO molecules into the inner areas of compact polymer particles. In other words, due to the PEO attachment the hydrophilicity of the thermally-collapsed precursor increases, therefore, the conditions for its further modification more and more approximate to the conditions of modification of the precursor in the well-swollen coil state, that is, to the conditions for “random” modification of a chain that existed for the same reaction performed in solution, i.e. at 15 °C in this case. Hence, when carrying out surface chemical hydrophilization of the macromolecular precursor that has been changed by any external stimulus to the collapsed state, one must ensure the limited extent of modifications in order to inhibit gradual swelling of the interior areas of the globular core and prevent the reaction of all possible active groups (here, GMA) of the precursor chains.

Another serious problem related to the “chemical colouring” approach for the synthetic preparation of protein-like copolymers is the working concentrations of the reaction participants [macromolecular precursor and low-molecular-weight (or oligomeric) modifying agent]. The point is that in the case of the theoretical design of such macromolecules, each of them is considered to be an individual copolymeric chain, and the procedure of “computer colouring” of the surface of a dense globule is carried out on a single particle, which does not interact physically with other polymeric partners [1–3]. For a real chemical reaction leading to the formation of a protein-like core-shell structure, this means the necessity to work in dilute solutions of macromolecular precursors, at least at a concentration of less than $C^* = 1/[\eta]$, in order to ensure the stimulus-induced formation and subsequent surface modification of individual globules. However, the efficacy of chemical processes at low reagent concentrations is, as a rule, rather poor (slow reaction rate, small yield of target products). Therefore, one has to maintain a certain compromise between the real chemical opportunities and maintaining as low as possible the concentration of reagents affording folded globular conformation of a high polymer precursor.

Nonetheless, one cannot exclude the probability of a successful combination of these prerequisites (as was the case with poly[(NiPAAm-*co*-GMA)-*g*-PEO considered above]) that will allow us to obtain, using the “chemical colouring” approach, the protein-like ***HP***-copolymers with a dense hydrophobic core wrapped by the hydrophilic shell. Such a shell should be capable of efficiently protecting the temperature-responsive macromolecules against pronounced interchain hydrophobic interactions and precipitation at temperatures significantly higher than those at which the copolymers of the same total monomer composition—but with a non-protein-like primary sequence of comonomer units—are in the soluble state.

2.2.2 Partial Hydrophobization of Hydrophilic Polymer Precursors

In this section we will touch upon an alternative route for the synthesis of protein-like copolymers using the “chemical colouring” approach, i.e., the approach mentioned in Sect. 2.1, viz. partial hydrophobization of the initially hydrophilic macromolecular precursors. In this case, it is also of great significance to guarantee the action of the forces that "force" the *de novo* hydrophobized segments of chains to assemble together into the core. Evidently, for the process within an aqueous media these forces are the hydrophobic interactions that can be strengthened by the elevated temperatures, that is, the modification reaction should be carried out over the temperature range that is the most favourable for hydrophobic interactions.

Recently, such a procedure has been used by Wu and Shanks [22] upon a preparation of partially hydrophobized polyacrylamide (**PAAm**). In this work, the transamidation reaction has been employed to convert amide groups of the parent polymer to the alkylamide groups of the modified PAAm. Scheme 2 shows the reaction chart for the case of the *N*-isopropylamide derivative.

It was found that in spite of the large excess of modifying amine (*N*-isopropyl-, -diethyl, -dipropyl, -diisopropyl, -*n*-hexyl, -cyclohexyl, -*n*-octyl), the extent of substitution did not exceed 5–10 molar %. For the case of the *N*-isopropyl derivative, i.e. [poly(AAm-*co*-NiPAAm)], the authors connected such results with the temperature-induced conformational transformation of partially hydrophobized copolymer acquiring the contracted conformation, "... which made it difficult for *N*-isopropylamine to react further with the amide groups" [22]. Unfortunately, no data on the solution behaviour of these interesting copolymers have been reported to date, although there is a high probability that they would demonstrate certain properties of the protein-like macromolecules. At least, in favour of similar supposition is supported by the results of our studies [23] of somewhat different PAAm partially hydrophobized derivative, whose preparation method is depicted in Scheme 3.

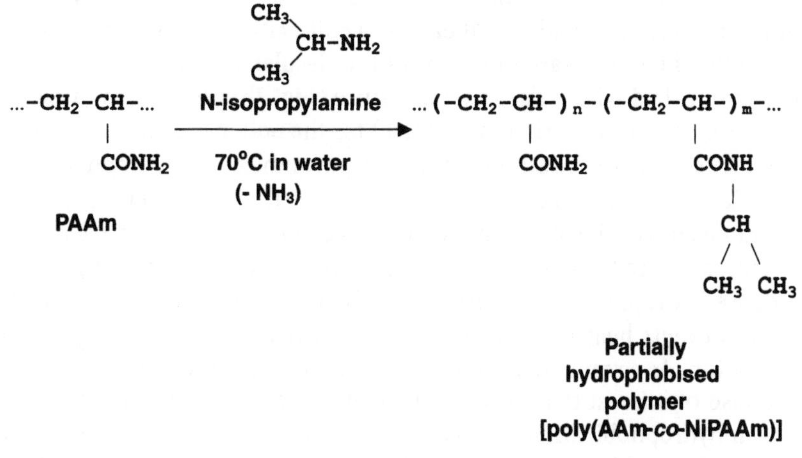

Scheme 2

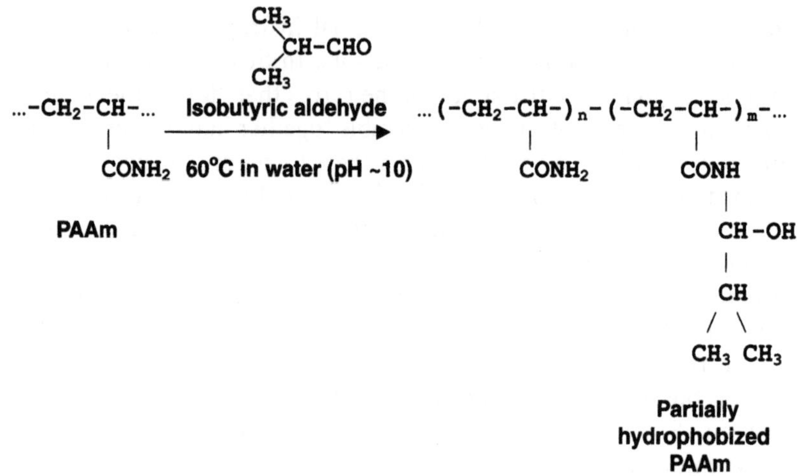

Scheme 3

This synthesis was based on the known hydroxyalkylation reaction of unsubstituted amides with aldehydes [24]. The modification of PAAm with isobutyric aldehyde was performed in a moderately alkaline aqueous medium (pH $\sim$ 10) at 60 °C, i.e., at a temperature higher than LCST for PNiPAAm (the modified units are close structural analogues of the NiPAAm units). Solutions of the resulting partially hydrophobized PAAm (according to NMR data, the extent of substitution was about 7 mol %) in pure water during capillary viscometry studies at 20 and 50 °C showed differences in comparison with solutions of the initial PAAm (Fig. 2). The values of reduced and intrinsic viscosities of PAAm dilute solutions increased with the rise in temperature (Fig. 2a), showing the upswelling phenomena for PAAm coils owing to the enhancement of affinity in between the polymer and the solvent. At the same time, the viscosities of the solutions of partially hydrophobized derivative were, firstly, lower (that is, this derivative had a smaller size of coils, although chemical modification resulted in an increase in the overall molecular weight) and, second, the values diminished with a rise in temperature (Fig. 2b) pointing to the shrinking phenomena (decrease in particle size) for the polymer coils. The latter fact was obviously due to the strengthening of intramolecular hydrophobic interactions. Therefore, it can be supposed that above the critical temperature, the macromolecules of this partially hydrophobized PAAm derivative decrease in size. This is due to the conformational transition giving rise to the formation of a dense hydrophobic core surrounded by the hydrophilic shell. The shell, in turn, prevents heat-induced intermolecular hydrophobic aggregation and precipitation. As a result, the copolymer remains in the soluble state.

It is also of interest that the behaviour of the reaction system in the course of PAAm hydrophobization with isobutyric aldehyde also testifies to the conformational transition as the chemical modification proceeds. After addition

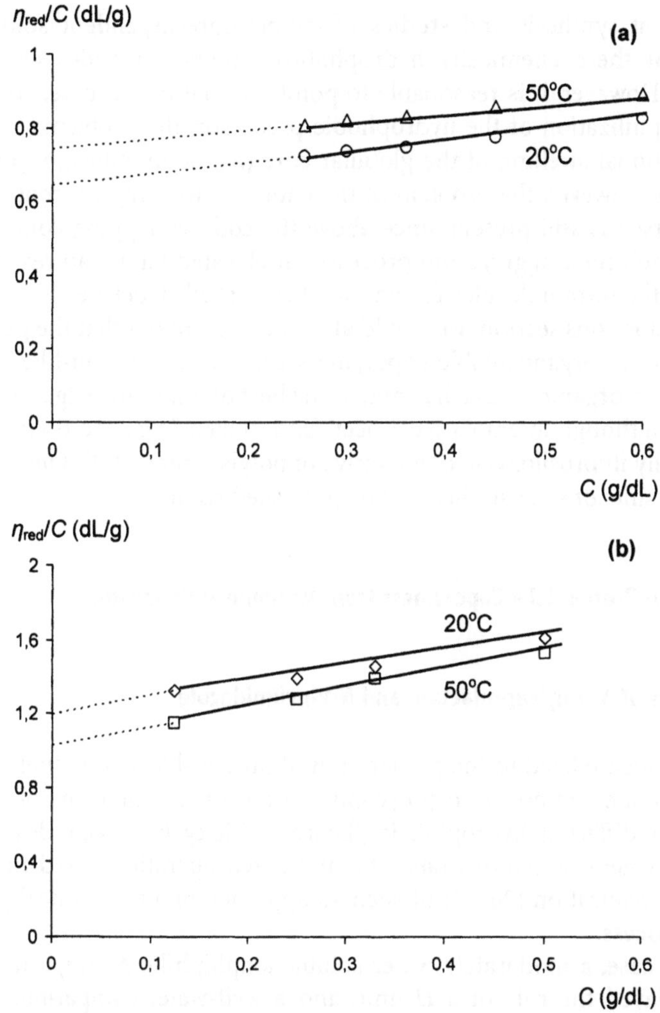

Fig. 2 **a** Concentration dependences of reduced viscosity values at 20 and 50 °C for aqueous solutions of PAAm and **b** partially hydrophobized PAAm prepared accordingly to Scheme 3 (the data from [23])

of the aldehyde to the polymer solution the system becomes opaque rather quickly, thus pointing to the partial loss of water-solubility by the resulting modified polymer. Then, about 15 minutes later, the solution again transforms to a transparent fluid, thus, by implication, pointing to a change in chain conformation. It can be supposed that in this case, because of the hydrophobic interaction of "novel" pendant groups, an inversion of the initial conformation occurred. This gave rise to the formation of a more hydrophobic compact core surrounded by the non-modified hydrophilic (or only slightly modified) shell.

Efforts at synthesis and studies of temperature-dependent solution behaviour of these chemically hydrophobized polyacrylamides are now in progress. However, it is reasonable to point out that in this case, contrary to the hydrophilization of the hydrophobic precursor, the problems associated with additional swelling of the globular core (as the modification proceeds) are absent; however, the problem of the choice of working concentration for the precursor is still present since above the coil overlapping concentration the intermolecular aggregation processes at elevated temperatures can compete with the intramolecular formation of core-shell structures.

As we close this section, we would also like to point out that the purposeful preparation of organosoluble copolymers exhibiting "protein-like" solution behaviour in organic media, has not, to the best of our knowledge, so far been reported, although attempts have been undertaken (e.g., the preparation of the partially fluorosiloxylated derivative of polyisoprene [25]). One may hope that the results of such studies will be published soon.

2.3 Synthesis of Protein-Like Copolymers from Monomeric Precursors

2.3.1 Copolymers of *N*-Vinylcaprolactam and *N*-Vinylimidazole

A general idea related to the preparation of protein-like copolymers through the co-polymerization or co-polycondensation of the mixtures of comonomers with differing hydrophilicity/hydrophobicity has been described in Sect. 2.1. Scheme 4 demonstrates the multi-step operations used in the first successful realization [26, 27] of such an approach in a free radical polymerization process.

In this case, a moderately water-soluble amphiphilic *N*-vinylcaprolactam (**NVCl**) played the role of a ***H***-unit, and a well-water-compatible *N*-vinylimidazole (**NVIAz**) served as a ***P***-unit. The polymerization was carried out in a medium of 10% aqueous dimethylsulfoxide (**DMSO**). The addition of DMSO to the reaction solvent was necessary because of insufficient NVCl solubility in pure water. It was also shown that in this solvent mixture, the NVCl-homopolymers and NVCl/NVIAz-copolymers retained their LCST-behaviour [26, 28]. Hence, the DMSO in the reaction solvent did not significantly suppress the hydrophobic interactions of the NVCl units. The polymerization was initiated by the redox system (*N*,*N*,*N*′,*N*′-tetramethylethylenediamine (**TMEDA**) + ammonium persulphate (**APS**)) and was carried out at 65 °C (1st step). This condition was very important, since admittedly the temperature was higher than the phase separation threshold of the reaction bulk when the polymeric products were formed; that is, under these thermal conditions, hydrophobically-induced folding as the NVCl-blocks appear was ensured. After completion of the reaction, the

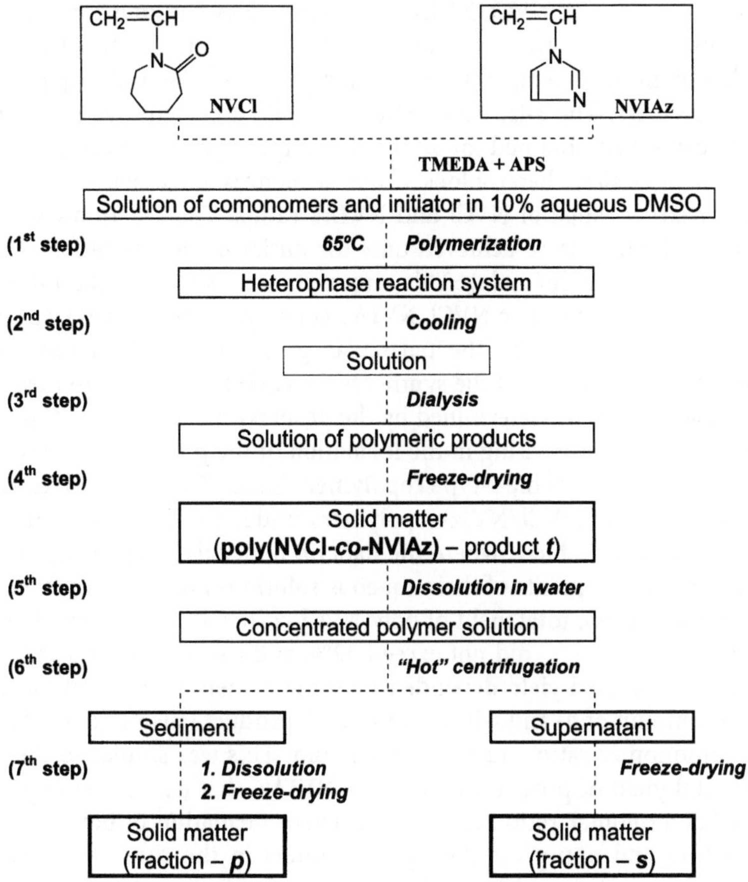

Scheme 4

opaque system was cooled down to room temperature (2nd step). This resulted in the formation of a transparent solution, which was then subjected to dialysis (3rd step) for removal of low-molecular-weight admixtures and oligomeric compounds. Such a procedure, instead of the usually employed precipitation of synthesized macromolecular products with an appropriate organic liquid, was used in order to operate in the aqueous medium only and does not change the conformation of copolymer chains acquired by them in water. After dialysis, the copolymer was freeze-dried (4th step) and then it was re-dissolved in water at necessary concentration (5th step). The next principal point of the preparative scheme under discussion was the "hot" separation stage (6th step). This technique allowed division of the total (*t*) polymeric products, i.e., poly(NVCl-*co*-NVIAz), into two different (in respect of their solution behaviour at temperatures higher than PST) fractions: a thermally precipitating (*p*) fraction and a thermally non-precipitating or soluble (*s*) fraction.

The latter kind of NVCl/NVIAz-copolymer was the one for which the protein-like structure was supposed, since the fraction ***s*** restored its water-solubility at elevated temperatures, as was predicted theoretically [1–3] for such copolymers. The ***s***-fraction, which exhibits a protein-like solution behaviour, cannot be obtained for all the initial ratios of the comonomers. The formation of desired heteroblock chain sequences took place only within a rather narrow range of NVCl and NVIAz ratios. The reaction conditions shown in Scheme 4 were achieved only for initial monomer ratios of about 85 : 15 to 90 : 10 (mole/mole). This circumstance, as well as the rather low chemical yield of respective NVCl/NVIAz-copolymers (see below) was, without doubt, conditioned by the particular properties of the given pair of comonomers. The results of the syntheses depended on the reactivities of the comonomers and were determined by the properties of each type of growing macroradical: those resulting in the formation of the ***p***-type macromolecules and those giving rise to the ***s***-type copolymer chains. The summarized results of the syntheses of NVCl/NVIz-copolymers under the conditions indicated above are given in Table 1, where the data on molecular weights of the fractions and the cloud points of their aqueous solutions are also presented.

On the whole, the total yield of poly(NVCl-*co*-NVIAz) in the precipitation polymerization at 65 °C did not exceed 32%, and the process was characterized by a bell-shaped yield dependence on the content of NVCl in the initial feed. The amount of ***p***- and ***s***-fractions in each product ***t*** also depended on the initial comonomer ratio and had a maximum. This was similar to the trend for the total yield of products ***t*** vs. NVCl/NVIAz, that is, a certain optimum of initial comonomer ratio existed for efficient formation of both thermally precipitating and non-precipitating copolymers in the course of precipitation copolymerization of NVCl/NVIAz mixtures at temperatures higher than PST. The data on monomer compositions of the isolated fractions ***p*** and ***s*** were of considerable interest, since the NVCl/NVIAz ratios found for these copolymers reflected, within definite limits, the efficiency of NVCl and NVIAz participation in the copolymerization. Thus, the highest amount of NVCl units and, hence, the lowest amount of NVIAz units (67 to 33 molar %, respectively) was found in the ***s***-copolymer formed from the 80 : 20 (mole/mole) NVCl/NVIAz initial mixture. That is, a marked enrichment of the thermally non-precipitating fraction with a hydrophilic monomer compared to its initial content in a feed took place. Among the ***p***-fractions the highest extent of enrichment with NVIAz units was also found to occur at the same initial comonomer ratio, but the enrichment was less (27.3 molar %) than that for the ***s***-fraction. These results pointed to the higher reactivity of NVIAz as compared to NVCl in aqueous media during the precipitation polymerization under the conditions used, i.e. at a temperature above PST.

In this respect, most interesting was the fact that the comonomer ratio found for the fractions ***p*** and ***s*** of the copolymer synthesized at the initial NVCl/NVIAz molar ratio equal to 85 : 15 turned out to be virtually the same,

Table 1 Results of copolymerization of NVCl with NVIAz under the reaction conditions indicated in Scheme 4 (the data from [27])

Initial comonomer ratio (mole/mole)	Fraction [a]	Yields of total product (%)	Fractions of the total product	Found comonomer ratio [b] (mole/mole)	Molecular weight [c]	Cloud point of 1 g/L aqueous solution (°C)
75 : 25	*t*	12.3	–	–	–	–
	p		65.9	78.0 : 22.0	225 000	44
	s		34.1	73.0 : 27.0	160 000	65
80 : 20	*t*	17.0	–	–	–	–
	p		65.4	72.7 : 27.3	150 000	35
	s		34.6	67.0 : 33.0	85 000	66
85 : 15	*t* + *o*	32.1	–	–	–	–
	o	39.5	–	81.0 : 19.0	270 000	33
	t	60.5	–	–	–	–
	p		69.2	74.0 : 26.0	105 000	34
	s		30.8	73.0 : 27.0	40 000	> 70
90 : 10	*t*	10.3	–	–	–	–
	p		67.5	88.0 : 12.0	390 000	37
	s		32.5	81.0 : 19.0	160 000	64

[a] *o* – oily deposit onto the walls of reaction vessel.
[b] Determined from the ^1H NMR spectra; experimental accuracy was ±10% from a value measured.
[c] The values correspond to effluent volumes of size exclusion chromatography peaks (column with CL-Sepharose 2B resin; eluent – 0.2 M NaCl; temperature – 20 °C).

namely, 74 : 26 and 73 : 27, respectively (Table 1). In turn, the fractions had significant differences in molecular weights and, importantly, exhibited absolutely different temperature-dependent solution behaviour. The fraction ***p*** began to precipitate at 34 °C (in fact, similar to the NVCl-homopolymers [28–31]), and the ***s***-fraction remained water-soluble at least up to 70 °C. The latter fraction, as well as the ***s***-fraction of copolymer obtained from the feed with an initial comonomer molar ratio of 90 : 10, are both marked in grey in Table 1 because they were identified to possess a protein-like structure in aqueous media at definite temperatures.

As to the molecular characteristics of the NVCl/NVIAz-copolymers, all the ***p***-fractions had higher molecular weights than the ***s***-fractions. This trend was the same as that frequently observed in precipitation polymerization, where the molecular weights of the resulting polymers are higher than for the analogous polymers synthesized in a solution (the main reason is a decrease in probability of chain growth termination in the recombination and disproportioning reactions due to decreasing mobility of macroradicals in the precipitating polymer phase) [32]. Differences in molecular weights [determined using size-exclusion chromatography (SEC)] probably affected the ability of different fractions of ***p*** to precipitate in a heated aqueous medium, because the comonomer composition of the samples did not differ drastically (Table 1).

Conformational and phase transitions can potentially be indicative of the primary structure of thermosensitive macromolecules. Indeed, depending on the relative location of ***H***- and ***P***-blocks, as well as on the variation of their length, the chains can either undergo conformational transition accompanied by phase separation, or they can exhibit only the conformational changes without macroscopic phase transitions, i.e. the behaviour observed in the case of protein-like ***HP***-copolymers. Therefore, the solution behaviour of separated fractions of these NVCl/NVIAz-copolymers in an aqueous medium at different temperatures is very important.

The last column in Table 1 shows the values of cloud point temperatures (T_{cp}) determined for these fractions. All the ***p***-fractions of NVCl/NVIAz-copolymers exhibited pronounced phase separation effects upon heating of their aqueous solutions. The turbidity of the systems increases sharply after certain threshold temperatures. This indicates the temperature-dependent phase separation typical for LCST-exhibiting polymers such as the NVCl-homopolymers [28–31]. At the same time, the ***s***-fractions of poly(NVCl-*co*-NVIAz) either did not precipitate from the aqueous solutions at all till 70 °C, or the LCST effects were observed at temperatures considerably higher (by 20–30 °C) than for the corresponding ***p***-fractions. Examples of thermonephelometric curves that demonstrate the temperature dependences of relative intensities of light scattering of the corresponding fractions of copolymers synthesized from the feeds with initial comonomer ratios of 85 : 15 and 90 : 10 (mole/mole) are shown in Fig. 3a and b, respectively.

Also, the figures contain nephelometric curves for the product ***t*** (that is, for ***p*** + ***s***, in the proportions formed in the synthesis) and the fraction ***o***, which was also a thermally precipitating product and deposited onto the walls of the reaction vessel in the course of the copolymerization of NVCl with NVIAz at their initial molar ratio of 85 : 15 (Table 1). One can see that the precipitation behaviour of the total product ***t*** differs, although the amount of the ***s***-fraction is almost the same at 31–33%. Obviously, this depends on the "properties" of the ***s***-fraction. For instance, the heat-induced precipitation of the sample ***t*** formed from the feed with a comonomer molar ratio of 90 : 10 (Fig. 3b) is suppressed by the presence of "its own" ***s***-fraction to a markedly lesser extent when compared to the product ***t*** obtained at the comonomer molar ratio of 85 : 15 (Fig. 3a). Most likely, such differences reflect the divergent influence of the ***s***-fractions on the coagulation processes in the thermo-precipitating fractions of the total product ***t***. These differences, for example different surface

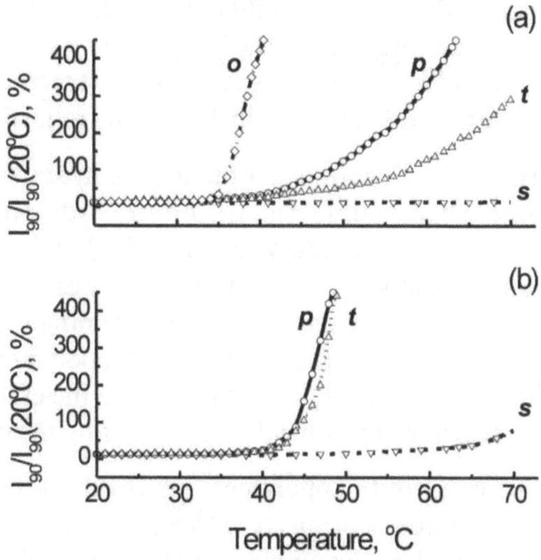

Fig. 3 Thermonephelometry curves of aqueous solutions of the ***o***-, ***t***-, ***p***- and ***s***-fractions of poly(NVCl-*co*-NVIAz) synthesized at 65 °C from the feeds with initial molar ratios of comonomers equal to **a** 85 : 15 and **b** 90 : 10 (the data from [27])

hydrophilicity of the *s*-type coils, can be due to the specific primary structure and, therefore, different spatial packing of the hydrophobic and hydrophilic blocks.

Among the copolymers of the *s*-type not all of them remained almost transparent upon heating up to 70 °C (curve *s* in Fig. 3b and the T_{cp} value for this copolymer in Table 1 correspond to this case), in spite of the fact that their separation from the thermally precipitating ***p***-fractions and isolation were both performed at 65 °C (Scheme 4). However, the increase detected in light scattering from the solution of this *s*-fraction at the temperatures of the nephelometric experiments was rather small when compared to that registered for the samples capable of pronounced heat-induced phase separation (***o***- and ***p***-fractions). In other words, the temperature-dependent solution behaviour of the ***p***- and *s*-fractions was very distinct. However, we would like to emphasize once again, their integral comonomer compositions differed rather weakly, especially in the case of the fractions prepared at initial NVCl/NVIAz molar ratio equal to 85 : 15 (Table 1). This means that the ***p***- and *s*- fractions, which were simultaneously formed "in one pot" during the course of the polymerization, have a rather different comonomer distribution along the chains. Because of this, the location of hydrophilic NVIAz units and their blocks in the macromolecules of the *s*-type ensured prevention of heat-induced macroscopic phase separation. Therefore, the attempts at finding other (besides the solubility at elevated temperatures) protein-like properties were focused (by the authors of the study [27]) on the *s*-fractions, as it was supposed that protein-like copolymers would be found among such thermally non-precipitating fractions.

Despite the fact that the water-solubility of the *s*-fractions during the heating of their aqueous solutions (above the PST for the *p*-fraction) is an essential property of protein-like macromolecules, other structural features could also be responsible for such behaviour, including the following:

- Random distribution of a significant number of hydrophilic NVIAz units along the polymer chain could result in uniform hydrophilization. This, in turn, could lead to a loss of ability for the coil-globule transition, which is caused by the hydrophobic interactions. As a result, such copolymers should be water-soluble over a wide temperature range.
- In the primary structure of alternating block-copolymers the hydrophobic ***H***-type blocks could be too short to facilitate the efficient cooperative hydrophobic interactions responsible for promoting the phase separation.

Therefore, in order to identify copolymers possessing a protein-like structure, it is necessary to choose a property with which one can distinguish them unambiguously from random copolymers or alternating block-copolymers.

One such property, as has been demonstrated (see [26]), is the change in partial heat capacity of the copolymer solution upon the heat-induced conformational transition of macromolecules. Such a change was detected by high-sensitivity differential scanning calorimetry (HS-DCS). The DSC data for the NVCl/NVIAz-copolymers synthesized at initial comonomer ratios of 85 : 15 and 90 : 10 (mole/mole) are given as thermograms in Fig. 4.

In general, the functions of partial heat capacity of the *p*-fractions has a similar profile typical of polymer solutions with LCST. The heat capacity curves go through a maximum within the temperature range of phase separation of the system. These peaks of heat capacity differ significantly from that of a pulse-like function, which, according to theoretical concepts, should describe the heat capacity change in a solution under phase separation. The heat capacity peaks are highly diffused. In most cases, they do not display a well-defined end point of the transition. For this reason, it was difficult to obtain reliable data on the transition enthalpy. However, some tendencies in the enthalpy change were found, depending on the synthesis conditions and on the composition of the isolated copolymer fractions.

The transition enthalpy decreased regularly with a decrease in the content of hydrophobic NVCL units in the copolymers. This trend is rather obvious, since, in general, the value of the transition enthalpy can be regarded as a measure of the cooperativity of hydrophobic interaction of the oligoNVCl-blocks: the higher the cooperativity, the larger the transition enthalpy. On the other hand, it is evident that the cooperativity of interaction of the blocks is determined by their length. On this basis, it is possible to evaluate changes in the length of thermoresponsive blocks in *p*-fractions by examining the enthalpy of phase transition. Apparently, the average length of such blocks decreases with an increase in NVIAz units. In this regard, the calorimetric data for the *s*-fractions of copolymers obtained from the feeds with initial

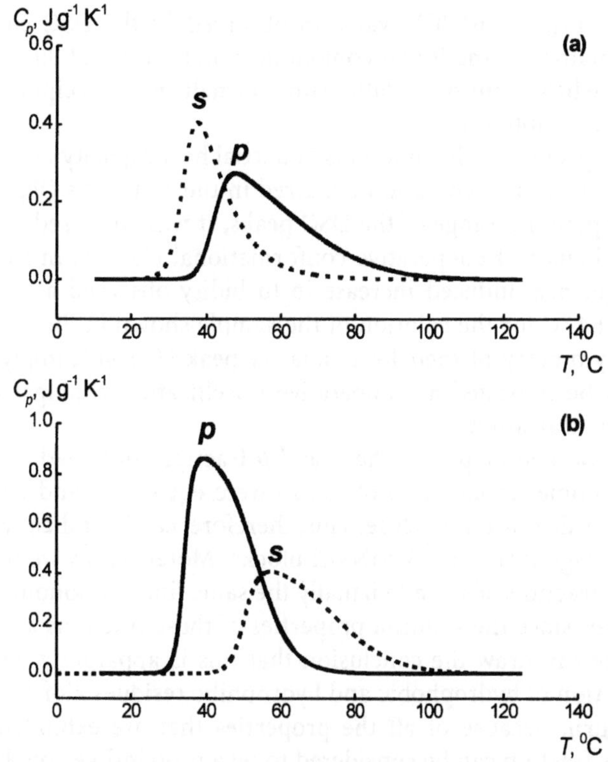

Fig. 4 Partial heat capacity functions of poly(NVCl-*co*-NVIAz) synthesized at 65 °C from the feeds with initial molar ratios of comonomers equal to **a** 85 : 15 and **b** 90 : 10 (the data from [26, 27, 42])

comonomer molar ratios of 75 : 25 and 80 : 20 (the data for these samples are not shown in Fig. 4) testified that such copolymers did not undergo any cooperative conformational changes upon heating. It is highly probable that this result reflects the insufficient length of the hydrophobic oligoNVCl-blocks. In fact, at a uniform (random) distribution of the NVIAz units along the chains in the *s*-fractions of the NVCl/NVIAz-copolymers with the comonomer molar ratios of 73 : 27 and 67 : 33 (Table 1), the length of the oligoNVCl-blocks would vary from ∼ 2 to ∼ 3 units, only. It is clear that NVCl-containing sequences that are so short cannot exhibit cooperative conformational changes. On the other hand, the *s*-fraction of the copolymer obtained from the feed with an initial comonomer ratio of 85 : 15 (mole/mole) also contained the comonomers in a proportion of 73 : 27 (similar to the discussed *s*-fraction formed at the initial comonomer molar ratio of 75 : 25), but showed the DSC-detected cooperative transition at $T_{max} \sim 37\,°C$ (Fig. 4a), thus indicating significant differences in the conformations of these two thermally non-precipitating NVCl/NVIAz-copolymers, most likely due to the difference in their primary sequences. Lastly, a very similar transition, but at higher temperatures ($T_{max} \sim 61\,°C$), was also observed in the DSC curve for the *s*-fraction obtained at the initial comonomer molar ratio of 90 : 10 (Fig. 4b). Therefore, the behaviour of the latter two thermally non-precipitating copolymers is of special interest.

For these specimens, the functions of partial heat capacity revealed a maximum. Since no phase separation occurred in these systems (Figs. 3a and b) over the temperature range of the DSC peaks, it was supposed that the DSC data gave evidence of cooperative conformational changes in these copolymers. A slight heat-induced increase in turbidity observed at temperatures higher than 64 °C for the solution of the sample shown in Fig. 3b, as well as a specific asymmetry of their heat capacity peak (Fig. 4b), imply that these changes can be indicated as cooperative micellization conjugated with the "coil-globule" transition.

The transition enthalpies of the *s*- and *p*-fractions obtained from the feed with a comonomer molar ratio of 85 : 15 were equal to 6 and 7 J/g, respectively, i.e. the values are very close. This, therefore, can be indicative of almost the same average length of oligoNVCl blocks. Moreover, as we have already stressed, the fractions also had virtually the same final comonomer composition. However, since the solution properties of these fractions are drastically different, one can draw the conclusion that this is apparently due to a specific distribution of hydrophobic and hydrophilic residues along the polymer chains. In turn, because of all the properties that are exhibited by the *s*-fraction, this fraction can be considered to be a protein-like copolymer [27].

The enthalpy of conformational transition of the *s*-fraction obtained from the feed with a comonomer molar ratio of 90 : 10 was 11 J/g, which is about a third of the enthalpy (27.3 J/g) of phase transition ($T_{max} = 41.8\,°C$) of the *p*-fraction. Hence, such a *s*-copolymer differs from the corresponding *p*-fraction through a shorter average length of hydrophobic oligoNVCl-blocks. This is also in agreement with the higher average content of hydrophilic NVIAz units in the *s*-fraction (Table 1). The HS-DSC-detected transition in the aqueous solution of the given *s*-type copolymer started at $\sim 47\,°C$ and showed T_{max} at 61.3 °C (Fig. 4b), whereas the cloud point of the system was observed at around 64 °C (Table 1); that is, on the descending branch of the DSC-curve. These facts can be indicative of the protein-like properties of such a *s*-fraction. At the very least, such a behaviour is characteristic of this copolymer over a temperature range from the onset of the HS-DSC peak to T_{cp}.

Thus, the results of the studies of poly(NVCl-*co*-NVIAz) synthesized in an aqueous medium at a temperature above the PST demonstrated that some of the copolymers exhibit typical LCST-properties, whereas other copolymers remain water-soluble upon the heating of their solutions but show the transition effects detected by HS-DSC (as well as by NMR and light scattering, see below). From these data, the following questions can be formulated:

(a) What are the main features of the primary structure of the ***o***-, ***p***- and ***s***-fractions of the copolymers under discussion?
(b) What are the possible reasons for the simultaneous formation of so distinct fractions in the course of the polymerization reaction?
(c) Which of the copolymers, among the synthesized and studied fractions, can be classified as protein-like copolymers?

Let us consider these problems step by step in more detail.

(a) The actual molar ratios of NVCl and NVIAz units in the copolymers of ***o***- and ***p***-types presented in Table 1 have varied from 78 : 22 to 88 : 12. Therefore, assuming a random distribution of the comonomers along the chains, the length of hydrophobic oligoNVCl blocks should be from ∼ 3 to ∼ 7 units. Such a length is too short to cause pronounced heat-induced phase separation effects typical for the thermoresponsive NVCl-copolymers. However, the phase transition for such thermally precipitating fractions started at the temperatures 33–44 °C, i.e., in the region of T_{cp} values characteristic of the NVCl-homopolymers (32–45 °C, depending on the polymer molecular weight [28, 30, 31, 33, 34]). Therefore, it can be assumed that all the NVCl/NVIAz-copolymers of the ***o***- and ***p***-types were not the random copolymers with the short oligoNVCl segments. They did possess a primary structure of the block copolymers, and the length of some of the NVCl-containing blocks was large enough to cause the pronounced LCST behaviour of such macromolecules. At the same time, the distribution of hydrophilic NVIAz units and oligoNVIAz blocks in the polymer chains was such that hydrophilic segments were unable to "shield" the expanded NVCl-rich parts against the intermolecular hydrophobic interactions upon heating. In other words, each macromolecule of the ***o***- and ***p***-type should contain at least several long enough "non-protected" oligoNVCl blocks.
A slight increase in the turbidity upon heating of aqueous solutions of the ***s***-fractions of the NVCl/NVIAz-copolymers obtained from the feeds with initial comonomer molar ratios of 75 : 25 (T_{cp} 65 °C) and 80 : 20 (T_{cp} 66 °C) could be due to the micellization phenomena, although the absence of DSC peaks over the same temperature range testified to the non-cooperative character of the process. This could indicate that the chains of these ***s***-type copolymers had, nevertheless, a certain amount of oligoNVCl blocks non-buried by the hydrophilic microenvironment sufficiently well and thus capable of participating in the hydrophobically-induced associative intermolecular processes at elevated temperatures. At the same time, the sequence of monomer units in the ***s***-copolymers obtained from the feeds with the initial comonomer ratios of 85 : 15 and 90 : 10 (mole/mole) corresponded to the block-copolymers of another type. The basis for such a conclusion is the lack of macroscopic heat-induced phase separation at elevated temperatures (Fig. 3 a and b) and, simultaneously, the transi-

tion effects revealed by the HS-DSC (Fig. 4a and b). This means that the primary structure of these specimens included, apart from the irregular sequences of comonomer units, both long enough hydrophobic oligoNVCl blocks and the long enough hydrophilic oligoNVIAz blocks. The former were responsible for the temperature-dependent cooperative transitions within the macromolecular coils, while the latter were capable of protecting the chains against the intermolecular hydrophobic aggregation. According to the computer simulation data [1–3], exactly such a primary structure should be characteristic of the protein-like ***HP***-copolymers.

(b) Another important problem concerns the reasons for the parallel formation of the thermally precipitating fractions, that is, ***p*** and ***o***, along with fraction ***s*** during the copolymerization of NVCl and NVIAz in an aqueous medium at temperatures above the PST. Indeed, in the course of copolymerization of NVCl and NVIAz under such thermal conditions, a strong attraction between the hydrophobic entities is promoted, which, therefore, should favour the formation of thermally precipitating copolymers of the ***o***- or ***p***-types. However, what mechanisms can be responsible for the simultaneous formation of the ***s***-fraction?
Most likely, one of the significant factors for this process is the instant location of the reactive end of a growing macroradical during the polymerization process. Possible variants of such a location can be as follows:

- either in a dissolved propagating chain, which is in a coil conformation, or;
- in a microheterogeneous system, where the growing end is located either inside, or at the surface of the dense hydrophobic core, and such a core is formed (due to the coil-globule transition) from an excessive amount of the more hydrophobic comonomer, NVCl, already at the early stages of precipitation polymerization.

The overall comonomer ratio, as well as the sequence of comonomer units along the chains of synthesized macromolecules, should obviously be different in these two cases. The chains of those thermally non-precipitating ***s***-fractions of the NVCl/NVIAz-copolymers that did not show conformational transition effects over the temperature range till, at least, 65 °C (the samples obtained from the feeds with the comonomer molar ratios of 75 : 25 and 80 : 20, Table 1) more probably started to grow after the reaction of the initiating radical and NVIAz. Thus, the propagation of more hydrophilic chains with a predominantly random (statistical) distribution of comonomer units occurs as predicted for the classical solution copolymerization [35, 36]. On the contrary, those reactive macroradicals, which start to grow from the mainly NVCl-containing sequences, acquire the conformation of collapsed chains, and their growing ends are located within the hydrophobic interior of the globules. These ends should predominantly react with NVCl molecules (this was shown for the case of

NVCl homopolymerization under the same conditions [28] and was also demonstrated in the computer experiments [5,7]). Thus, the formation of long enough oligoNVCl segments is facilitated, i.e., such a mechanism favours the formation of the ***o***- and ***p***-type copolymers.

Finally, the formation of a heterogeneous blockiness characteristic of the protein-like copolymers can be realized upon the fast formation of a sufficiently dense NVCl-enriched hydrophobic core and the subsequent "exit" of the growing end onto the surface of this collapsed small globule, where the hydrophilic NVIAz molecules can easily react. The elevated temperature is the factor maintaining the hydrophobic interior of such chains in a collapsed state, preventing the macromolecules (at least, their dense cores) from acquiring loose conformation as the copolymerization of the comonomers with different hydrophobicity/hydrophilicity occurs. Apparently, this is the only variant of the polymer chain growth capable of forming thermally non-precipitating fractions of the ***s***-type with the "hydrophobic core–hydrophilic shell" structure inherent in the protein-like ***HP***-copolymers.

Evidently, there are a number of factors that determine which of the above-mentioned pathways of polymerization occurs. One of these is the enrichment of final copolymers with the more hydrophilic comonomer (in comparison with the composition of the initial feeds), due to a higher reactivity of NVIAz. The higher (than for NVCl) reactivity of NVIAz is determined by a stronger polarization of the double bond of the vinyl group by the imidazole moiety as compared with the caprolactam cycle. Furthermore, a good solubility of hydrophilic NVIAz can also contribute to its higher reactivity in aqueous media.

(c) Theory predicts that the main feature of the primary structure of synthetic (non-natural) protein-like copolymers is a sequence of heterogeneous blocks possessing distinct relative hydrophilicity/hydrophobicity [1–8]. As was already pointed out, with a "favourable" combination of such blocks, a specific spatial packing of the macromolecular coils occurs in the aqueous medium: the dense core (enriched with ***H***-type monomer units) surrounded by a shell enriched with ***P***-type polar comonomer. If the oligo ***H*** blocks are long enough to exhibit the LCST-behaviour, an increase in temperature higher than a certain critical point should cause a conformational transition (collapse) of these blocks. However, highly hydrophilic oligoNVIAz blocks on the periphery of the collapsed core of protein-like copolymers should presumably prevent the inner regions from the intermolecular hydrophobic contacts, thus "preserving" such copolymers against the temperature-induced phase separation. Hence, only the combination of properties, namely, the temperature-induced conformational transition with a parallel retention of water solubility is a reliable indication of a protein-like ***HP***-copolymer.

In fact, among the products listed in Table 1 such behaviour observed upon heating of aqueous polymer solutions was detected only for the ***s***-fractions of NVCl/NVIAz-copolymers synthesized from the feeds with the initial comonomer ratios of 85 : 15 and 90 : 10 (mole/mole) (the data marked in grey in Table 1). The "indicators" of conformational transition were the changes in the heat capacities of the solutions detected by HS-DSC (Fig. 4a and b) in combination with a transparency (Fig. 3a) or a very slight turbidity (Fig. 3b) of the solutions at temperatures corresponding to the DSC peaks. Apparently, for the copolymer synthesized from the feed with the 90 : 10 comonomer molar ratio, a micellization took place upon heating of the solution above its T_{cp}, i.e., 64 °C (Table 1), when a slight increase in light scattering coinciding with the diffuse descending branch of the heat capacity curve. However, the onset of this HS-DSC peak was registered at a temperature about 17 degrees lower ($\sim$ 47 °C; Fig. 4b), when the polymer solution still remained transparent (Fig. 3b). So, the behaviour of this ***s***-fraction in the aqueous solution over the temperature range from 47 °C to T_{cp} (i.e., prior to micellization) can be characterized as the behaviour typical of the protein-like copolymers. Finally, the most pronounced effects were observed for the ***s***-fraction of poly(NVCl-*co*-NVIAz) prepared from the feed with a molar comonomer ratio of 85 : 15. Its aqueous solution remained transparent upon heating up to at least 70 °C (Fig. 3a), whereas endothermic transition was clearly observed by HS-DSC at temperatures from $\sim$ 30 to $\sim$ 60 °C (T_{max} 37.3 °C; Fig. 4a). Thus, the copolymer can be attributed to the protein-like copolymers. In this case, at elevated temperatures, the macromolecules should acquire the specific chain conformation typical of such protein-like entities. In the pre-transition temperature region, such macromolecules possess a less-dense interior, but the pre-requisites for their compactization (folding) with a rise in temperature already exist. Such pre-requisites can lead to intramolecular microsegregation of the blocks of different hydrophilicity/hydrophobicity.

Apart from the data of thermonephelometry and HS-DSC, ^{1}H NMR studies have also revealed [27] some properties that allowed the attribution of such ***s***-type copolymers to the protein-like ones. A marked broadening of the water proton signal was observed caused by the decreased mobility of bound water just in the vicinity of the temperature of HS-DSC peak. These data indicated the heat-induced compaction of the interior of the polymer coils, as would occur with protein-like macromolecules. Figure 5 demonstrates the experimental data, viz., the temperature dependences of signal width at half-height for the peaks of water protons recorded in D_2O-solutions of ***p***- and ***s***-fractions of the copolymer synthesized from the feed with an initial comonomer ratio of 85 : 15 (mole/mole).

It is known [37, 38] that conformational and phase transitions of water-soluble temperature-responsive polymers significantly influence the molecular dynamics of water molecules. Therefore, the studies of temperature dependence of the parameters capable of reflecting such mobility are of significant importance. Since the ^{1}H NMR signal width at half-height ($\Delta_{1/2}\nu$) of the spectral line of the water protons is inversely proportional to the spin-spin relaxation time T_2 (i.e., $\Delta_{1/2}\nu \sim 1/T_2$) [39], the $\Delta_{1/2}\nu$ parameter can characterize the molecular mobility. That is why this parameter was measured in the studies of NVCl/NVIAz-copolymers [27].

In the case of polymer solutions, the simplest model describing the motion of water protons is a two-phase model implying the co-existence of low mobility protons (the protons of the polymer-bound water molecules) and of high mobility protons (the protons of the bulk water molecules). The water protons are exchanged between these "phases" with the rate of $k_c = \tau_c^{-1}$, where τ_c is the correlation time. In turn, the width of NMR lines of highly mobile protons depends on the ratio of the correlation time and the frequency of spectrum registration (ω). If the proton exchange rate is small in comparison with the ω value, $\omega\tau_c \gg 1$, only the NMR signal of bulk water is observed, because the signal of bound water protons has very low intensity. Upon heating, the exchange rate increases, and τ_c decreases ($\omega\tau_c \sim 1$) resulting in a broadening of the spectral lines of the water protons. Further rise of temperature leads to $\omega\tau_c \ll 1$, and the value of $\Delta_{1/2}\nu$ becomes smaller, i.e., the line becomes narrower.

The values of $\Delta_{1/2}\nu$ for water protons in the D_2O-solutions of the ***s***- and ***p***-fractions under discussion were measured over a temperature range from 20 to 55 °C [27]. The heat-induced variations of the $\Delta_{1/2}\nu$ parameter for these two copolymers are different (Fig. 5a and b). For instance, at temperatures from 20 to 24 °C, the line width of the water proton signal in the solution of the fraction ***p*** (Fig. 5a) broadens from 1.88 to 3.5 Hz, it narrows further

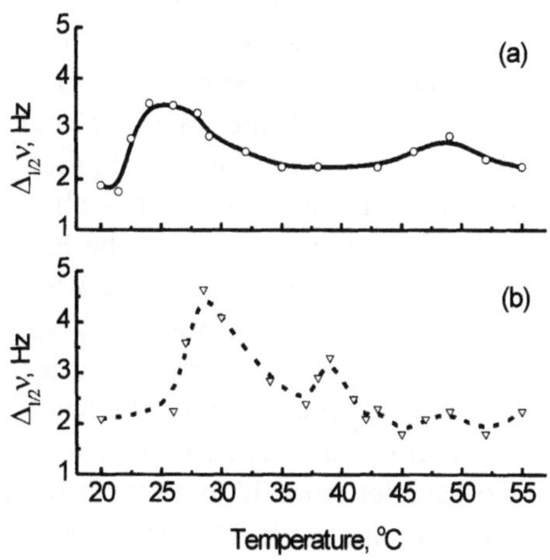

Fig. 5 Temperature dependences of the full width of NMR signal ($\Delta_{1/2}\nu$) at half maximum measured for water protons in D_2O solutions of the fractions **a** ***p*** and **b** ***s*** of poly(NVCl-*co*-NVIAz) synthesized at 65 °C from the feed with initial ratio of comonomers 85 : 15 (mole/mole) (the data from [27])

to 2.25 at 34–35 °C, is unchanged up to 43 °C, rises again a little to 2.85 Hz at 49 °C, and then diminishes to 2.25 Hz at 55 °C. As the cloud point of the *p*-type copolymer is around 34 °C (Table 1) commencing the phase separation effect (Fig. 3a), and the HS-DSC experiments detect the onset of the peak above the T_{cp}, these facts indicate that an increase in the $\Delta_{1/2}\nu$ value over the temperature range of 20–34 °C is mainly connected with changes of proton exchange rate between the bound and bulk water. At the same time, the broadening of the water proton signal at temperatures 43–49 °C was evidently caused by the phase separation in the system, when the polymer precipitates. Hence, the mobility of bound protons considerably decreased, and the rate of proton exchange decelerated. At further heating, due to increasing mobility of bulk water molecules, only the signals of their protons were registered (the condition of $\omega\tau_c \ll 1$). This was manifested in the descending branch of the curve in Fig. 5a at temperatures from 49 to 55 °C. Interestingly, a very similar kind of variation in water proton mobility with temperature was observed [40] in the D_2O-solutions of another thermally precipitating polymer, NiPAAm-homopolymer, when the T_2 values were measured. In that case, more pronounced effects were registered. Certainly, this was due to the homopolymeric nature of the PNiPAAm macromolecules and the high cooperativity of the phase transition.

At the same time, the temperature-dependent variations of the $\Delta_{1/2}\nu$ parameter for the solution of fraction ***s*** were different as compared to the fraction ***p***. The variation of the width of the NMR signal of the water protons (Fig. 5b) was even more pronounced with increasing temperature as compared to the variation for the ***p***-fraction (Fig. 5a). The first peak on the curve for the ***s***-fraction is higher and was detected at a higher temperature. The mobility of water protons in polymer solutions depends on the conformation of the macromolecular chains (e.g., an average density of the polymer coil). Therefore, such a "temperature shift" testifies that for the ***s***-type copolymer there was a stronger interaction with the water molecules than for fraction ***p***, because the condition $\omega\tau_c \sim 1$ is reached at a higher temperature for the ***s***-fraction. The second peak on the curve in Fig. 5b is observed over the temperature range of the descending branch of the HS-DSC peak (Fig. 4b). Inasmuch as the conformational transition of the chains of thermoresponsive polymers in an aqueous medium is accompanied by a collapse of such macromolecules [31, 41], such a collapse has to result in a decrease in mobility of bound water and, thereby, in a broadening of the whole width of the NMR signal from the water protons. This was observed for the ***s***-copolymer as an increase in the $\Delta_{1/2}\nu$ parameter from 37 to 39 °C. It indicates a compaction of the structure of these macromolecules at the temperatures, where the transition effect was registered by the HS-DSC, but no macroscopic phase separation took place. On further heating the proton mobility increased, and this gave rise to a narrowing of the respective NMR signal, whose line width varied insignificantly at least till 55 °C. Therefore, the NMR data on such heating-induced compaction of the interior of these *s*-type (i.e. soluble at elevated temperatures) macromolecules also gave proof of their protein-like structure.

Also, it should be noted that since D_2O was the solvent employed in the NMR experiments instead of H_2O that was used in the optical and calorimetric studies, a possible difference in the temperatures has to be taken into account, when comparing the results of the NMR studies with the effects observed by thermonephelometry or HS-DSC.

In addition, data on the size, shape and solvation of the polymer particles in aqueous solutions at temperatures below and above the transition phenomena registered by HS-DSC have been obtained [42]. Table 2 shows the results of capillary viscometry and light scattering experiments for the fractions ***p*** and ***s*** of poly(NVCl-*co*-NVIAz) synthesized at 65 °C from the feed with the initial molar comonomer ratio equal to 85 : 15. Since fraction ***p*** precipitates from the aqueous solution at temperatures > 34 °C, its intrinsic viscosity can be determined only at 20 °C, whereas for the fraction ***s*** such measurements were possible above and below the temperatures of the HS-DSC-registered conformational transition.

In this study, it was found that the data of the viscosity measurements obeyed well the linear dependences in standard coordinates "reduced viscosity–polymer concentration", and the coefficients of determination were close to 1. Therefore, the differences revealed between the $[\eta]$ values at 20 and 50 °C can be considered as reliable. These values *per se* were quite small, especially for the fraction *s*, thus obviously testifying to the rather compact conformation of such macromolecules in water at both temperatures. At the same time, the studies revealed a pronounced association of these copolymers in aqueous medium even at 20 °C, when both solutions looked transparent and no conformational transitions were detected. Thus, the about 13-fold higher value of M_w as compared to M_w found in the DMSO solution for the fraction ***s*** testified to the presence of aggregates. The value of the conformation-sensitive parameter $\rho = R_G/R_h$ close to 1 indicated the spherical shape of the polymer particles [43], and the negative value of the second virial coefficient showed that water at 20 °C was not a thermodynamically good solvent for this copolymer. The aggregation extent depended on the fraction type and the temperature of measurements. At 20 °C, about 7–8 chains of ***p***-type copolymer in the dilute aqueous solution combined into aggregates. For the ***s***-fraction, this value was equal to 12–13. With an increase in temperature and strengthening of hydrophobic interactions, the aggregation increased 6–8 times.

However, certain contradictions can be seen from the data of Tables 1 and 2. Indeed, the molecular weights determined for aqueous solutions of these ***p***- and ***s***-fractions in the SEC experiments (Table 1) coincided well with the results of light scattering for DMSO solutions (Table 2); but why did the molecular weights differ so considerably from the light scattering data (Table 2) for the solutions of given copolymers in pure water, where, in

Table 2 Molecular parameters of the thermally precipitating and non-precipitating fractions of poly(NVCl-*co*-NVIAz) [a] in aqueous solutions at 20 and 50 °C (the data from [42])

Fraction	T (°C)	$[\eta]$ (dL/g)	M_w ($\times 10^{-3}$) [b]	$A_2\ 10^4$ (cm^3 mole g^{-2})	R_G (nm)	R_h (nm)	$\rho = R_G/R_h$	$d = (M_w/N_A)\upsilon^{-1}$ 10^3 (g/cm^3) [c]	$1/d$ (cm^3/g)
P	20	0.102 ± 0.009	1890 (254)	−4	197	119	1,7	0,1	10 000
	50	–	14 240	−12	144	121	1,2	1,9	526
S	20	0.064 ± 0.003	380 (30)	−18	210	228	0,9	0,02	50 000
	50	0.057 ± 0.002	2270	9	220	217	1,0	0,1	10 000

[a] Obtained from the feed with initial comonomer molar ratio equal to 85 : 15.
[b] The value in parentheses is the average-weight molecular weight found in static laser light scattering experiments for DMSO solution of the copolymer.
[c] $\upsilon = (4/3)\pi R_G^3$

spite of pronounced aggregation, the intrinsic viscosities were found to be so small?

First of all, it is of importance to understand weather other—rather than hydrophobic intermolecular interactions—can be responsible for the formation of large water-soluble aggregates at the temperatures below transition detected by HS-DSC and NMR. For instance, donor-acceptor interactions between the carbonyl carbon atom (reduced electron density) of the caprolactam cycle and non-sheared electron pair of the nitrogen atom in the imidazole cycle could be one of the possible additional types of interaction. It is clear that such interactions are weak, but with increasing NVIAz content they should strengthen. It is also known [44] that predisposition to such polar interactions increases, if the interacting groups are partially surrounded by non-polar groups preventing whole hydration of the polar ones, and especially, if an association of the numerous non-polar groups occurs. Certainly, in the case of poly(NVCl-*co*-NVIAz) the latter condition is fulfilled, because the amount of NVCl units is 2–4 times higher than NVIAz units. Furthermore, at temperatures below the LCST, the absence of large aggregates in aqueous solutions of NVCl homopolymer (in which the polar interactions under consideration do not present) was demonstrated in the light scattering experiments [45]. This obviously testifies to the high probability of the participation of such donor-acceptor interactions between the NVCl and NVIAz units in the formation of soluble aggregates of the copolymers. Besides, such interactions should be sensitive to the presence of low-molecular-weight electrolytes, and probably such aggregates were not registered in the gel-chromatography experiments, when molecular weights of the copolymer fractions were determined in the medium of 0.2 M NaCl (see footnote (c) in Table 1).

Secondly, not only the size of polymer particles, but also their actual density and shape in aqueous solutions should be taken into account for the explanation of the contradictions between the small $[\eta]$ and gross M_w values. Thus, the more asymmetric shape ($\rho > 1$, Table 2) of the aggregates consisting of the ***p***-type chains (as compared with the shape of aggregates consisting of the ***s***-type macromolecules) can, as is already known [44], contribute, along with the chain length, to the higher $[\eta]$ value for fraction ***p***. In addition, the influence of polymer solvation on the viscosity characteristics is seen from Table 2 upon comparison with the samples with a close shape ($\rho \approx 1$), but with a distinct affinity to the solvent ($A_2 > 0$ and $A_2 < 0$), as in the case of fraction ***s*** at 20 and 50 °C. It can be supposed that at a high thermodynamic affinity of polymer particles to a solvent ($A_2 > 0$; ***s***-fraction at 50 °C), the latter one is held within the particles and moves together with them. In that case, the particle can be approximated by an equivalent solid sphere [44]. On the contrary, at poor affinity to a solvent ($A_2 < 0$; ***s***-fraction at 20 °C), that is, when the latter one is held by the polymer particles only weakly, the model of a free draining particle is more feasible for the description of its hydrodynamic properties [44]. From the Kratky plot, it was found [42] that the behaviour of the *s*-fraction at 20 °C in water can be described as intermediate between the behaviour of solid spheres and worm-like chains. The calculations also showed for this case a very low density ($d = 0.02$; Table 2) of aggregates consisting of such chains. This parameter grew significantly at 50 °C, which manifests a compaction of the particles. Also, light scattering studies showed that below the temperature of the HS-DCS-detected transition, water is not a good solvent for the macromolecules of fraction *s*, whereas above the transition temperature, the thermodynamic affinity between the copolymer and water increases. Certainly, this is because of the change in conformation, when more hydrophobic segments are "concentrated" in the interior, while the hydrophilic ones are on the periphery of the polymer particles. One can say that in this case, these macromolecules acquire the conformation inherent in the globular proteins.

When comparing the $[\eta]$ values with the reciprocal values of density of polymer particles, $1/d$, calculated for the copolymer samples from the light scattering data, one can see (Table 2) about two-fold differences of intrinsic viscosities of the *p*-fraction at 20 °C and *s*-fraction at 50 °C, whereas the $1/d$ values were analogous for these copolymers. Under such thermal conditions, the affinity to water was so different that A_2 values had inverse signs. This, in turn, shows that for the most complete explanation of hydrodynamic properties of the *p*- and *s*-fractions, such characteristics of the copolymers as density and shape of the particles (individual macromolecules or their soluble aggregates/polymeric micelles) should be taken into account, in combination with the degree of solvation. In the case of the protein-like fraction *s*, this indicates that the data of both viscometry and light scattering studies unambiguously reveal the temperature-induced shrinking of polymeric particles, because their density increases, and intrinsic viscosity decreases.

All the described properties of such a *s*-fraction of poly(NVCl-*co*-NVIAz) synthesized at the temperature above the PST of the reacting system allowed us to draw the conclusion that the chains of this type had the comonomer sequence, which at the temperatures above the conformation transition facilitated the formation of polymer particles, where ***H***-blocks are in the interior shielded by the ***P***-blocks against additional intermolecular association. Such a behaviour of this copolymer in aqueous media is close to that of oligomeric proteins similar to casein [46] possessing a rather hydrophobic core surrounded by the polar segments.

2.3.2
Copolymers Containing *N*-Isopropylacrylamide-Based Blocks

The polymers and some copolymers of NiPAAm are well-known [41, 47–54] to possess a brightly-expressed LCST behaviour in aqueous media. In recent years this behaviour has attracted growing interest because of the possible applications of such macromolecular systems in diverse areas, especially in the field of Life Sciences [55–57]. As for the synthesis of protein-like copolymers using NiPAAm as the basis for the creation of stimuli-sensitive sequences, such examples were also described.

Scheme 5 depicts the copolymerization of NiPAAm and NVIAz, which was used [58] for the preparation of so-called polymeric displacers employed in immobilised metal chelate chromatography (**IMCC**) of proteins [59]. This synthesis was carried out at room temperature from a feed with an initial comonomer ratio of 10 : 1 (mole/mole). Isolation of total (***t***) polymeric products from the reaction bulk was accomplished by their thermal precipitation at 75 °C. Thereafter, the obtained poly(NiPAAm-*co*-NVIAz) was separated into the two fractions using IMCC on iminodiacetate-sepharose loaded with Cu^{2+}-ions (**Cu^{2+}-IDA-sepharose**) capable of binding oligoNVIAz sequences via the formation of chelate complexes. Upon passing the polymer solution through the column, the breakthrough effluent contained the unbound fraction of the copolymer ($M_W = 1.4 \times 10^6$), which included ∼ 7.8 mol % of NVIAz units. Further rinsing of the column with 0.02 M EDTA eluted the Cu^{2+}-IDA-sepharose-bound fraction ($M_W = 1.35 \times 10^5$) with a somewhat higher (9.1 mol %) content of NVIAz.

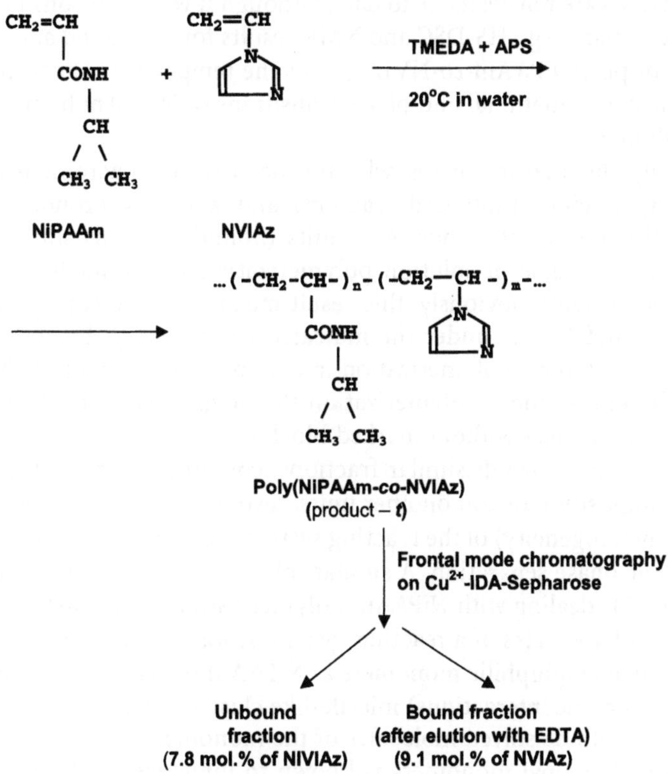

Scheme 5

These results show that in a water medium, even at synthesis temperatures below PST, the copolymerization of comonomers possessing non-equal hydrophilicity/hydrophobicity enables simultaneous generation of copolymers with significantly different properties. In this study, it was supposed that the copolymer, which did not interact with the metal chelate adsorbent (the unbound fraction), had a virtually random distribution of NVIAz units along the chains and did not contain sufficient amounts of the long enough oligoNVIAz blocks. On the other hand, specifically the resin absorbed the copolymer possessing the protein-like sequence of the monomer units, where the hydrophilic pendant imidazole groups of the oligoNVIAz blocks (which are responsible for the complexing ability) were accumulated at the hydrophilic periphery of the macromolecular coil. The IMCC-behaviour can be explained because of the structure and conformation of these bound and unbound fractions. Neither the small differences in NVIAz content, nor the differences in molecular weights of the fractions (both are able to penetrate into the sepharose matrix, so pronounced exclusion effects can hardly be the reason for the separation of unbound matter) could be responsible for the chromatographic distinctions observed.

Unfortunately, data on the temperature-dependent solution behaviour of these fractions are not available to date, although it will be of considerable interest to compare, e.g., HS-DSC and NMR results for the bound and unbound fractions of poly(NiPAAm-*co*-NVIAz) over the temperature range characteristic of the conformational and phase transitions of NiPAAm homopolymers and copolymers.

Naturally, the question arises: why are such distinct macromolecules with most likely random (unbound fraction) and a clearly pronounced non-random distribution of comonomer units (bound fraction) simultaneously formed in the course of solution polymerization in the medium of a formally good solvent? Obviously, this result means that the copolymerization of NiPAAm and NVIAz under the indicated conditions (Scheme 5) was not a classical free-radical polymerization in a homogeneous solution, inasmuch as in the latter case the copolymerization should give rise to the formation of statistical copolymers without marked blockiness [35, 36]. The simultaneous formation of these two dissimilar fractions, combined with the very similar average composition of comonomer units, testifies to the microinhomogeneity (microheterogeneity) of the reacting system even below the temperature of macroscopic phase separation. A similar inhomogeneity was also mentioned in a study [53] dealing with NiPAAm polymerization in an aqueous medium at different phase states of a reacting system. It was assumed that in aqueous solutions such amphiphilic monomers as NiPAAM or NVCl tend to form (due to the hydrophobic interactions) micelle-like clusters, which are in a dynamic equilibrium with dissolved molecules of the monomer. A similar association of water-soluble vinyl monomers is known to influence markedly the polymerization of *N*-alkylacrylamides in aqueous media [50]. If the assumption about the microheterogeneity of the initial solution at 20–25 °C is valid for the copolymerization of the NiPAAm/NVIAz pair, the initiation of the polymerization process should result in the parallel formation of—at least—two types of chains. The first are those that started to grow and then propagate in a molecular-dispersed (dissolved) state. The second type of macromolecules are formed mainly from the monomers organized in clusters, thus memorizing, to a certain extent, their ordered structure. This mechanism can explain the presence of both ***p***- and ***s***-fractions in the composition of the final polymeric products ***t*** synthesized at temperatures below the PST for the case of NVCl/NVIAz-copolymers [26, 27], as well as the Cu^{2+}-IDA-sepharose-bound and unbound fractions for the case of NiPAAm/NVIAz-copolymers [58].

Interesting results on the properties of copolymers containing another pair of monomers, NiPAAm and *N*-vinylpyrrolidone (**NVP**), were obtained by the group of Chi Wu [54], who synthesized poly(NiPAAm-*co*-NVP) in an aqueous medium in solution at 30 °C and in suspension at 60 °C (i.e., higher than the PST) (Scheme 6). The redox initiating mixture TMEDA/potassium persulphate (**PPS**) was employed at both temperatures. The initial comonomer ratios were 95 : 5 or 90 : 10 (mole/mole). After the isolation and purification of the obtained copolymers, their monomer composition was determined with NMR spectroscopy, and respective aqueous solutions were investigated using light scattering and DSC methods.

It was shown that the final copolymers synthesized at 30 °C from the feeds with molar comonomer ratios of 95 : 5 and 90 : 10 contained 4.8 and 11.4 mol % of NVP units and had molecular weights 4.2×10^6 and 7.9×10^6, respectively. Poly(NiPAAm-*co*-NVP)s that formed in the course of precipita-

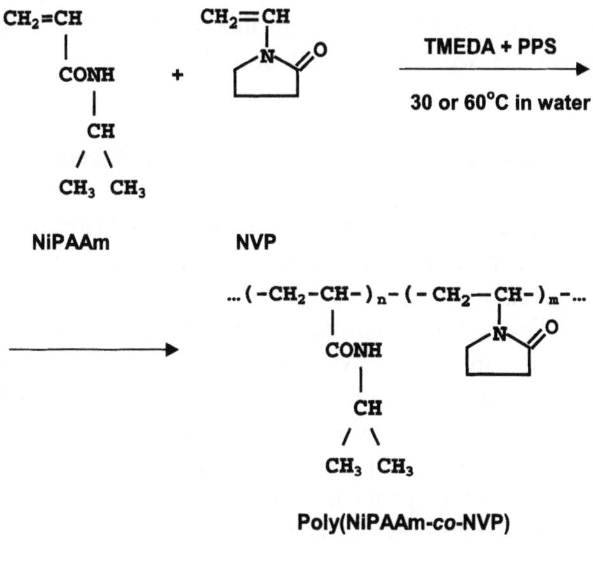

Scheme 6

tion polymerization at 60 °C contained 4.8 and 9.7 mol % of hydrophilic NVP units, respectively, and the molar weights of these copolymers were 2.9×10^6 and 5.6×10^6, respectively. In other words, the polymerization degrees of the copolymers prepared from the equal feeds under these two thermal conditions differed rather markedly, and the monomer composition of the copolymers differed insignificantly. However, it should be noted that these were the total (***t***) macromolecular products. No attempt was made to separate the synthesized copolymers into the thermally precipitating and non-precipitating fractions.

In regard to the problem of protein-like ***HP***-copolymers, the most significant data were obtained by the laser light scattering studies. It was revealed that copolymers synthesized above the PST had denser globular packing of macromolecules in solutions as compared to the copolymers fabricated from the same initial comonomer mixtures below the transition point. Thus, in dilute aqueous solutions the particle densities of poly(NiPAAm-*co*-NVP)s prepared at 30 and 60 °C from the feed with an initial comonomer ratio of 95 : 5 (mole/mole) were 0.070 and 0.096 g/cm^3, respectively, and for the copolymers produced from the feed of 90 : 10-composition the differences in the particle densities were well-pronounced: 0.033 and 0.073 g/cm^3, respectively. The authors drew the conclusion that such results were the consequence of the different distribution of monomer units along the chains. Namely, poly(NiPAAM-*co*-NVP), whose formation took place in the homogeneous medium at 30 °C, possessed a virtually random distribution of NVP units in the chains. On the other hand, the copolymer formed in the heterogeneous medium at elevated temperature, which facilitated chain folding, possessed the "protein-like segmented NVP distribution" [54]. Without doubt, for a more reliable validation of such a conclusion, additional studies of primary sequences of these protein-like copolymers, as well as all those discussed in Sect. 2.3, will be extremely desirable.

2.3.3 Synthesis of Protein-Like Copolymers Using Polycondensation Processes

When discussing various methods for the synthesis of protein-like ***HP***-copolymers from the monomeric precursors (Sect. 2.1), we pointed to the possibility of implementation of both polymerization and polycondensation processes. The studies of the potentials of the latter approach in the creation of protein-like macromolecular systems have already been started. The first published results show that using true selected reactions of the polycondensation type and appropriate synthetic conditions (structure and reactivity of comonomers, solvent, temperature, reagent concentration and comonomer ratio, the order of the reagents introduction into the feed, etc.) one has a chance to produce the polymer chains with a desirable set of monomer sequences.

For example, such attempts were undertaken in a body of work [60] dealing with the preparation (through acceptor-catalytic copolyesterification) of triple copolyesters composed from the residues of aromatic dicarboxylic acids, bisphenols and *N*-methyldiethanolamine (NMDEA). Scheme 7 illustrates the principle of such a syntheses: the upper part shows the chemical

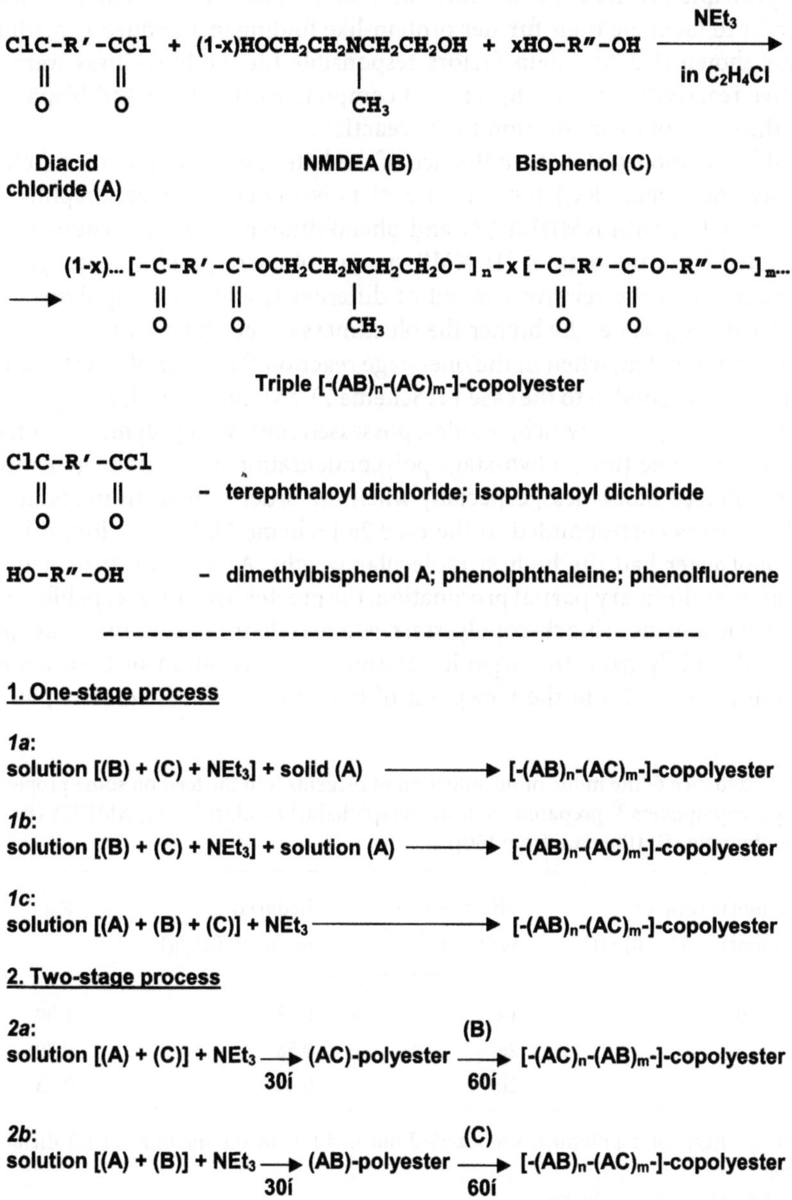

Scheme 7

structures of the comonomers; the bottom part shows the variants of reaction carried out and the order of introduction of the reagents into the feed.

By varying all the parameters of the process the authors prepared a set of copolyesters containing units and blocks with a tertiary amino group, which, in turn, could be transformed into a hydrophilic hydrochloride salt thus imparting water-compatibility to the initially organosoluble macromolecules. The principle involved was the formation of a block-type structure of chains in order to facilitate their further protein-like folding in an aqueous medium. It was shown that the main factors responsible for the blockiness were the relative reactivities of the (**B**) and (**C**) components (NMDEA and bisphenol) and the order of their addition to the reaction.

Table 3 demonstrates the influence of the latter factor on the microheterogeneity coefficient (K_M) for the case of polycondensation of terephthaloyl dichloride (**A**) with NMDEA (**B**) and phenolfluorene (**C**). This coefficient is computed from the data of ^{1}H NMR spectra of corresponding copolyesters. It characterizes the relative content of different triades in a copolymer: the smaller the K_M value, the higher the blockiness of the chains [61].

One can see that, when in the one-stage reaction the order of reagent introduction corresponded to the case 1a (Scheme 7), a virtually random copolymer was formed ($K_M \sim 1$), which, besides, possessed the lowest polymerization degree. At the same time, a two-stage polycondensation allowed the generation of pronounced blockiness, especially when the order of reagent introduction in the process corresponded to the case 2a (Scheme 7). In addition, this latter copolyester had the highest molecular weight. As a result, it turned out that after preliminary partial protonation, the greater hydration capability was inherent just in these block-copolyesters, whereas their random analogues precipitated rapidly from the organic solutions upon addition of even a small amount of water. From the viewpoint of the authors of this research, an in-

Table 3 Influence of the mode of introduction of reagents into the feed on some properties of triple copolyesters [a)] prepared from the terephthaloyl dichloride (**A**), NMDEA (**B**) and phenolfluorene (**C**) (the data from [60])

Initial molar ratio of comonomers (**A**) : (**B**) : (**C**)	The mode of synthesis [b)]	Reduced viscosity (dL/g) [c)]	K_M
0.5 : 1.0 : 0.5	1a	0.28	1.09
	2a	0.58	0.29
	2b	0.44	0.55

[a)] The synthesis of copolymers was carried out at 40 °C in the medium of 1,2-dichloroethane.

[b)] See bottom part of Scheme 7.

[c)] Measured at 25 °C for the tetrachloroethane solutions of copolymers.

creased stability of block-copolyesters in the aqueous media was promoted by the formation of a protein-like core-shell conformation of such macromolecules. Since more studies of these copolymers are in progress [62], one may hope for new experimental evidences supporting similar suppositions.

Generally speaking, we should note that the synthetic potentials of the polycondensation approach for the preparation of protein-like copolymers are only now beginning to be explored. Nonetheless, even the first results show promising directions for future synthetic developments that can hardly be reached through the polymerization processes.

For instance, one can imagine the following hypothetic scheme.

Let us assume that some protein-like ***HP***-copolymer (e.g. that derived theoretically via the computer simulation design [6]) has the following primary sequence of monomer units: ***HPPPHHPPPPHPPHHHHPHPPHHHHHPPPP PHH***... and so on (heterogeneous blockiness characteristic of the protein-like copolymers). In this case, the molar per cent of each block consisting of the units of one particular type (***H*** or ***P***) can be easily calculated, that is, the copolymer contains h_1% of ***H***, h_2% of ***HH***, h_3% of ***HHH***, h_4% of ***HHHH***... h_n% of $\boldsymbol{H_n}$, and also p_1% of ***P***, p_2% of ***PP***, p_3% of ***PPP***, p_4% of ***PPPPP***... p_n% of $\boldsymbol{P_n}$. Then, let us turn to possible chemistry. If there is such a set of blocks, they can be mixed in these proportions in an appropriate solvent, and after that the conditions need to be adjusted to the state capable of causing the association of hydrophobic ***H***-type blocks. Subsequent addition of a coupling agent will result in the polycondensation of the blocks with a certain probability of the formation of protein-like chains. Finally, the latter ones should be separated from other, non-protein-like, copolymers that appeared in the system.

Is such a synthesis possible?

It is thought, yes. For instance, if one type of the blocks has amino groups at both ends, another type –COOH-groups, and the blocks are water-soluble, the trigger for the association of more hydrophobic blocks can be the increase in temperature that facilitates the hydrophobic interactions. As a coupling agent, a water-soluble carbodiimide can be used. Then, after completion of the polycondensation, one can employ a procedure similar to "hot centrifugation" (Scheme 4) in order to separate the resulting polymeric products into the thermally precipitating and thermally non-precipitating fractions, and further to analyze the latter fraction for the presence of protein-like macromolecules. It is clear that in such a case, the ***H***-type blocks will react only with the ***P***-containing blocks. Therefore, the elongation of the blocks consisting of the chemically identical units will not occur. This, in turn, can result in the formation of desired sequences.

Certainly, other chemical structures (end groups) and suitable reactions can also be used in such a process. The main requirement of such "combinatory chemistry systems" is the conjunction of the initial solubility of the functionalized blocks in the solvent used and the existence of the "driving force" for the physical association of the more hydrophobic ***H***-blocks.

2.4
Successive Augmentation of Chains as a Way for the Preparation of Protein-Like Copolymers

This approach to the synthesis of protein-like *HP*-copolymers has not, as pointed out in Sect. 2.1, been used so far. Therefore, one may discuss only the potential opportunities of such a method for the creation of macromolecular chains with the desired protein-like sequences.

It is clear that the idea to explore the methodology of the solid-phase synthesis looks rather evident. At that, either successive augmentation of a chain step-by-step with each monomer unit accordingly to a pre-assigned program, or the attachment of certain separately synthesized blocks to a growing end, or a "mixed" variant (both single units and blocks coupling) can be implemented, as is usually employed in solid-phase peptide synthesis [10, 11, 63]. If the swelling ability of the growing chains in solvents used during the successive steps is maintained at a sufficient level, the coupling stages can be repeated many times. On the other hand, if the swelling ability decreases, the coupling efficacy decreases progressively due to chains folding and arising, as a result, additional difficulties for the reagents to interact with the "chemically active" growing end.

However, it is reasonable to emphasize that even in the case of the good swelling characteristics of the copolymer to be formed, the possible (achievable) length of the chains with correct primary sequences will be rather limited because of the accumulation of the "mistakes" owing to the non-quantitative yield of each synthetic stage—a well-known problem in the solid-phase synthesis of polypeptides or polynucleotides [10]. Although in the case of synthetic protein-like *HP*-copolymers this problem is not so drastic, as for the unique sequences of natural biopolymers (since for the synthetic macromolecules we always have a set of chains with various lengths and sequences), the accumulation of "incorrect" sequences in the growing chains should require us to carry out some separation operations with the final product in order to isolate the macromolecules with the desired combination of properties. Lastly, the time-, reagent- and solvent-consumptive character of the solid-phase synthesis should be noted, as well as the necessity to use expensive automatic synthesizers to produce long enough chain molecules. Therefore, if it is the aim of a researcher to obtain protein-like copolymers with the implementation of the given approach, the strategy of the synthesis, as well as the procedure of subsequent isolation of the target macromolecular product, have to be elaborated extremely thoroughly prior to the start of a project. Otherwise, the chances for a successful conclusion will be rather poor. Nonetheless, the prospect of obtaining some synthetic *HP*-copolymer with the exactly defined primary sequence is very attractive, and, therefore the possibilities of solid-phase synthesis should not be disregarded.

3
Conclusions

Interest in artificial (synthetic) protein-like copolymers has increased over the last few years, as such macromolecular systems are not only of theoretical (purely scientific) interest, but also of applied significance. For instance, on their basis, certain high molecular catalysts can be developed, in which a biomimetic type of chain packing, i.e. a dense hydrophobic core surrounded by a hydrophilic shell, will facilitate higher activity and specificity of such polymeric catalysts in comparison to other catalysts with the same chemical functions, but with random distribution along the chains. In this respect, the protein-like copolymers can be attributed to the popular group of so-called "smart" polymers. Therefore, elaboration of simple and efficient methods for the chemical synthesis of protein-like copolymers is of importance. Besides, an understanding of the mechanisms of formation of the protein-like macromolecules can help to better understand the diverse aspects of molecular evolution of biological macromolecules, since governed by the thermodynamic factors heterogenization of primary sequences of biopolymers composed of even two monomer units of differing hydrophilicity/hydrophobicity could, in the early stages of molecular evolution, result in the molecular selection of "evolutionally effective" structures and sequences. Recently, this aspect has been emphasized in a theoretical paper [64] and has been independently demonstrated in a series of impressive experimental works [65–67] devoted to the study of prebiotic synthesis of sequential peptides from the amino acid *N*-carboxyanhydrides under conditions modelling those of the early Earth. All of these facts allow us to consider the problems connected with the protein-like copolymers as significant not only for polymer chemists and physicists, but also for experts in other fields.

Acknowledgements The author thanks Professors A.R. Khokhlov, V.Ya. Grinberg and V.A. Vasnev for productive discussions, valuable advice and help during the review preparation. The INTAS (01-0607), Program for "Development and Study of Macromolecules and Macromolecular Structures of Novel Generations" (Russian Academy of Sciences) and Federal Target Scientific & Technical Program "R&D in Priority Directions of Science and Engineering" (Russian Ministry of Education and Science) are acknowledged for financial support.

References

1. Khokhlov AR, Khalatur PG (1998) Physica A (Amsterdam) 249:253
2. Khokhlov AR, Khalatur PG (1999) Phys Rev Lett 82:3456
3. Zheligovskaya EA, Khalatur PG, Khokhlov AR (1999) Phys Rev E 59:3071
4. Govorun EN, Ivanov VA, Khokhlov AR, Khalatur PG, Borovinsky AL, Grosberg AYu (2001) Phys Rev E 64:040903
5. Berezkin AV, Khalatur PG, Khokhlov AR (2003) J Chem Phys 118:8049

6. Khokhlov AR, Khalatur PG, Ivanov VA, Chertovich AV, Lazutin AA (2002) SIMU Newsletters Issue 4, Chap IV, p 79–100 (http://simu.ulb.ac.be/newsletters/newsletter.html)
7. Khokhlov AR, Khalatur PG (2004) Curr Opp Solid State & Mater Sci 8:3
8. Khokhlov AR, Berezkin AV, Khalatur PG (2004) J Polym Sci Part A Polym Chem 42:5339
9. Nilssen P, Hansen J, Ban N, Moore PB, Steitz TA (2000) Science 289:920
10. Hodge P, Sherrington DC (eds) (1980) Polymer-Supported Reactions in Organic Synthesis. Wiley, New York
11. Burgess K (ed) (2000) Solid-Phase Organic Synthesis. Wiley, New York
12. Garnet-Flaudy F, Freitag R (2000) J Polym Sci Part A Polym Chem 38:4218
13. Berezkin AV, Khalatur PG, Khokhlov AR, Reineker P (2004) New J Phys 6:44
14. Bo G, Wesslen B, Wesslen KB (1992) J Polym Sci Part A Polym Chem 30:1799
15. Gao G, Varshney SK, Wong S, Eisenberg A (1994) Macromolecules 27:7923
16. Choucair A, Eisenberg A (2003) J Am Chem Soc 125:11993
17. Yuk SH, Bae YH (1999) Crit Revs Ther Drug Carrier Syst 16:385
18. Discher DE, Eisenberg A (2002) Science 297:967
19. Virtanen J, Baron C, Tenhu H (2000) Macromolecules 33:336
20. Virtanen J, Tenhu H (2000) Macromolecules 33:5970
21. Virtanen J, Lemmetyinen H, Tenhu H (2001) Polymer 43:9487
22. Wu S, Shanks RA (2004) Polymer Int 53:1821
23. Lozinsky VI, Grinberg VY, Babushkina TA, Semenova MG, Khokhlov AR (2005) Proc Europ Polym Congr Moscow Russia i.4.2.1
24. Vail SL, Barker RH, Moran CM (1966) J Org Chem 31:1642
25. http://polymer.physik.uni-ulm.de/research/ReportOct2002/Report-Oct.2002.pdf
26. Lozinsky VI, Simenel IA, Kurskaya EA, Kulakova VK, Grinberg VY, Dubovik AS, Galaev IY, Mattiasson B, Khokhlov AR (2000) Doklady Chemistry 375:273
27. Lozinsky VI, Simenel IA, Kulakova VK, Kurskaya EA, Babushkina TA, Klimova TP, Burova TV, Dubovik AS, Grinberg VY, Galaev IY, Mattiasson B, Khokhlov AR (2003) Macromolecules 36:7308
28. Lozinsky VI, Simenel IA, Kurskaya EA, Kulakova VK, Galaev IY, Mattiasson B, Grinberg VY, Grinberg NV, Khokhlov AR (2000) Polymer 41:6507
29. Tager AA, Safronov SV, Berezyuk EA, Galaev IY (1994) Colloid Polymer Sci 272:1234
30. Kirsh YE, Yanul NA, Kalninsh KK (1999) Eur Polym J 35:305
31. Kirsh YE (1993) Progr Polym Sci 18:519
32. Arest-Yakubovich AA (1974) In: Encyclopaedia of Polymers. Sovetskya Entsiklopediya Publ, Moscow, 2:260 (in Russian)
33. Solomon OF, Corclovel M, Boghina C (1968) J Appl Polym Sci 12:1843
34. Maeda Y, Nakamura T, Ikeda I (2002) Macromolecules 35:217
35. Semchikov YD (1977) In: Encyclopedia of Polymers. Sovetskya Entsiklopediya Publ, Moscow, 3:446 (in Russian)
36. Tirrell DA (1985) In: Mark HF, Bikales NM, Overberger CG, Menges G, Kroschwitz JI (eds) Encyclopedia of Polymer Science and Engineering. Wiley, New York e.a. 4:192
37. Terada T, Inaba T (1994) Macromol Chem Phys 195:3261
38. Zeng F, Tong Z, Feng H (1997) Polymer 38:5539
39. Bailey RT, North AM, Pethrick RA (1981) Molecular Motion in High Polymers. Clarendron Press, Oxford
40. Ohta H, Ando I, Fujishige S, Kubota K (1991) J Polym Sci Part B Polym Phys 29:963
41. Schild HG (1992) Progr Polym Sci 17:163
42. Lozinsky VI, Simenel IA, Semenova VG, Belyakova LE, Il'in VV, Grinberg VY, Dubovik AS, Khokhlov AR (2006) Vysokomolekul soed 48(3) will appear (in Russian)

43. Burchard W (1994) In: Ross-Murphy SB (ed) Physical Techniques for the Study of Food Biopolymers. Blackie Academic and Professional, Glasgow, p 151
44. Tanford C (1961) Physical Chemistry of Macromolecules. Wiley, New York - London
45. Sibileva MA, Sibilev AI, Klyubin VV (2001) Vysokomolekul soed 43:1202 (in Russian)
46. Holt C (1992) Adv Protein Chem 43:63
47. Shibayama M, Tanaka T (1993) Adv Polym Sci 109:1
48. Wang X, Qiu X, Wu C (1998) Macromolecules 31:2972
49. Platé NA, Lebedeva TL, Valuev LI (1999) Polym J 31:21
50. Barabanova AI, Bune EV, Gromov AV, Gromov VF (2000) Eur Polym J 36:479
51. Larsson A, Kuckling D, Schönhoff M (2001) Coll Surf A Physicochem Eng Asp 190:185
52. Erbil C, Sarac AS (2002) Eur Polym J 38:1305
53. Lozinsky VI, Kalinina EV, Putilina OI, Kulakova VK, Kurskaya EA, Dubovik AS, Grinberg VY (2002) Vysokomolekul soed 44A:1906 (in Russian)
54. Siu MH, Liu HY, Zhu XX, Wu C (2003) Macromolecules 36:2107
55. Kanazawa H, Matsushima Y, Okano T (1998) Trends Anal Chem 17:435
56. Galaev IY, Mattiasson B (1999) Trends Biotechn 17:335
57. Hoshino K, Taniguchi M (2002) In: Galaev I, Mattiasson B (eds) Smart Polymers for Bioseparation and Bioprocessing. Taylor & Francis, London, p 257
58. Wahlund PO, Galaev IY, Kazakov SA, Lozinsky VI, Mattiasson B (2002) Macromol Biosci 2:33
59. Kumar A, Galaev IY, Mattiasson B (2000) J Chromatogr B Biomed Sci Appl 741:103
60. Markova GD, Vasnev VA, Keshtov ML, Vinogradova SV, Garkusha OG (2004) Vysokomolekul soed 46A:615 (in Russian)
61. Slonim IYa, Urman YaG (1982) NMR Spectroscopy of Heterochain Polymers. Khimiya, Moscow, p 87 (in Russian)
62. Keshtov ML, Markova GD, Vasnev VA, Khokhlov AR (2005) Vysokomolekul soed 476A:725 (in Russian)
63. Steward JM, Young JD (1969) Solid Phase Peptide Synthesis. W.H. Freeman & Co., San Francisco
64. Chertovich AV, Govorun EN, Ivanov VA, Khalatur PG, Khokhlov AR (2004) Eur Phys J–E 13:15
65. Commeyras A, Collet H, Boiteau L, Taillades J, Vandenabeele-Trambouze O, Cottet H, Biron JF, Plasson R, Mion L, Lagrille O, Martin H, Selsis F, Dobrijevic M (2002) Polym Int 51:661
66. Boiteau L, Collet H, Lagrille O, Taillades J, Vayaboury W, Giani O, Schue F, Commeyras A (2002) Polym Int 51:1037
67. Lagrille O, Taillades J, Boiteau L, Commeyras A (2002) Eur J Org Chem 1026

Adv Polym Sci (2006) 196: 129–188
DOI 10.1007/12_054

Published online: 6 December 2005

Role of Physical Factors in the Process of Obtaining Copolymers

Semion I. Kuchanov · Alexei R. Khokhlov (✉)

Polymers & Crystals Chair, Physics Department, Moscow State University,
119992 Moscow, Russia
khokhlov@polly.phys.msu.ru

Abstract A brief introduction into the principles of the statistical description of the chemical structure of linear heteropolymers is given. A comprehensive analysis of diffusion-controlled polymeranalogous reactions is provided in terms of the statistical chemistry and the statistical physics of polymers. The interplay of these two statistical approaches is conclusively demonstrated for the discussion of the quantitative theory of free-radical copolymerization in anomalous systems. The impact of physical factors on principal regularities of interphase free-radical copolymerization has been discussed in detail.

Keywords Chemical structure of macromolecules · Free-radical copolymerization · Polymeranalogous reactions

Abbreviations

MSD	molecular structure distribution
SCD	size-composition distribution
PAR	polymeranalogous reaction
SSL	strong segregation limit
WSL	weak segregation limit
DMF	dimethyl formamide
SAI	surface-active initiator
l	chemical size of a macromolecule
ξ	composition of a macromolecule
$f^{c}(\xi)$	composition distribution
$W(l\|\xi)$	fractional composition distribution
σ_l^2	dispersion of distribution $W(l\|\xi)$
M_α	monomer of α-th type
M_α	concentration of monomer M_α
$\overline{\mathrm{M}}_\alpha$	monomeric unit of α-th type
$\overline{M}_\alpha$	concentration of units $\overline{\mathrm{M}}_\alpha$
$R_\alpha(l)$	concentration of α-th type radical with chemical size l
R_α	overall concentration of α-th type radicals
x_α	monomer mixture composition
X_α	instantaneous copolymer composition
$\langle X_\alpha \rangle$	average copolymer composition
T_g	glass transition temperature
$f_\alpha^{bl}(n)$	distribution of α-th type blocks for length n
$Y_{\alpha\beta}(n)$	two-point chemical correlator
$W_{\alpha\beta}(x)$	generating function of $Y_{\alpha\beta}(n)$
R	radius of globule
h	Thiele modulus
Z	low-molecular reagent
n^*	scale of decay of chemical correlations
Φ_g	volume fraction of all monomeric units in globule
$I(\theta)$	intensity of scattering at angle θ
χ	Flory-Huggins parameter of pair interactions
χ_{sp}	value of parameter χ at spinodal
T	temperature
φ_α	volume fraction of monomer M_α inside globule
y^α	volume fraction of α-th phase in heterophase reaction system
r_1, r_2	reactivity ratios
l_α	number of units $\overline{\mathrm{M}}_\alpha$ in macromolecule
I_α	rate of initiation reaction in α-th phase

κ_1, κ_2 thermodynamic parameters of interphase copolymerization model
V_α rate of copolymerization in α-th phase
$\bar{\eta}_\alpha$ average length of block of units $\overline{M}_\alpha$
$\bar{n}_\alpha$ average number of blocks of units M_α
$\bar{l}_\alpha$ average number of units $\overline{M}_\alpha$ in macromolecule
p overall conversion of monomers
x_1^*, x_2^* azeotropic composition of monomers

1 Introduction

A specific feature of the study of polymer systems is the necessity to take into account the interplay between physical and chemical factors. For instance, the appearance of the phase diagram of a heteropolymer melt or solution is prescribed by the chemical structure of its macromolecules being formed at the stage of the synthesis of the particular polymer specimen. In its turn, the statistical characteristics of this structure describing composition, sequence distribution and architecture of macromolecules obtained may be governed by physical effects taking place in the course of their synthesis. Hence, treating real polymers, a theorist is supposed to resort to the approaches of both statistical physics and statistical chemistry of macromolecules. These two branches of polymer theory are destined for the solution of different problems. The goal of statistical chemistry is the investigation of the dependence of the statistical characteristics of macromolecules on the kinetic parameters of a reaction system, whereas statistical physics is concerned with establishing correlations between these characteristics and physical properties of polymers.

The prime objective of this concise review is to provide an illustration of the interaction of these two disciplines using particular examples. In choosing the examples, we seek to demonstrate the potentialities of the conformation-dependent design of the sequences of monomeric units in heteropolymer macromolecules. Under such a design, their chemical structure is controlled not only by the kinetic parameters of a reaction system but also by the conformational statistics of polymer chains.

2 Statistical Description of the Chemical Structure of Linear Heteropolymers

Examples considered in this paper deal with binary heteropolymers prepared by the method of chemical modification of homopolymers and by the free-radical copolymerization of a mixture of two monomers, M_1 and M_2. The products of each of these processes is a mixture of an enormous (practically infinite) number of various macromolecules differing in number of the constituent monomeric units $\overline{M}_1 = A$ and $\overline{M}_2 = B$ as well as in the fashion of their arrangement. An exhaustive description of the chemical structure of such a stochastic copolymer implies the formulation of a certain algorithm, which permits finding the Molecular Structure Distribution (MSD), i.e., the probability for a macromolecule with any configuration to be found in a particular polymer specimen [1]. Mathematically speaking, this means specification of the probability measure on the set of sequences of two symbols corresponding to units A and B. Under such a treatment, every macromolecule is associated with a certain realization of a stochastic process R of conventional movement along a polymer chain. This stochastic process has two regular states, S_1 and S_2, according to the number of types of units, while the role of "time" is performed here by the distance from the beginning of the macromolecule. A trajectory abandoning its limits falls into absorbing state S_0 to remain there from this moment on. Such a stochastic process with discrete "time" and several regular states is referred to in mathematics as a stochastic chain. The best known among them is the Markov chain [2–4] for which the probability $\nu_{\alpha\beta}$ to fall at any step into state S_β whose type is β depends exclusively on the type α of the state S_α at the preceding step. A Markov chain is fully defined by the matrix of transitions $\boldsymbol{Q}$ with elements $\{\nu_{\alpha\beta}\}$ and the initial vector $\boldsymbol{\nu}$ whose component ν_α equals the probability for the trajectory to start at state S_α. In the particular case when all rows of matrix $\boldsymbol{Q}$ are identical, the Markov chain is reduced to the stochastic process of independent trails. A heteropolymer specimen whose chemical structure is described by such a stochastic process is known as a random heteropolymer. Markovian copolymers are not rare in practice, being products of traditional processes of the synthesis of high-molecular compounds [5].

Macroscopic properties of a polymer specimen are, to a great extent, controlled by some statistical characteristics of the chemical structure of the macromolecules involved. For heteropolymers, such characteristics may be conveniently divided into two types.

2.1 Size-Composition Distribution

The main statistical characteristic of the chemical structure of a heteropolymer among those pertaining to the first type is the distribution of molecules $f(l_1, l_2)$ for numbers l_1 and l_2 of their constituent monomeric units $\overline{M}_1$ and $\overline{M}_2$. In dealing with a high-molecular weight polymer, these numbers may be taken as continuous variables, uniquely specifying chemical size $l = l_1 + l_2$ and composition $\zeta = l_1/l$ of a macromolecule. Under such a consideration, it is more convenient instead of function $f(l_1, l_2)$ to use the equivalent function of Size-Composition Distribution (SCD) $f(l, \zeta)$. This is possible to represent as the product

$$f(l,\zeta) = f(l)\,W(l|\zeta) \qquad f^c(\zeta) = \int_1^\infty f(l,\zeta)\mathrm{d}l \tag{1}$$

of size distribution $f(l)$ and fractional composition distribution $W(l|\zeta)$ which are equal, respectively, to the fraction of size l macromolecules and the fraction among them of those having composition ζ. The composition distribution $f^c(\zeta)$ of a polymer specimen as a whole is obtained by integration of the SCD over all values of variable l. The knowledge of distribution $f(l,\zeta)$ is of prime importance, because applying a chromatographic technique it is possible to separate copolymer molecules according to their size and composition [6]. The first- and the second-order statistical moments of this SCD are of special importance

$$X \equiv \langle\zeta\rangle \qquad \sigma_l^2 \equiv \langle(\zeta - X)^2\rangle \, . \tag{2}$$

In these formulas the letter X stands for the average copolymer composition, while σ_l^2 denotes the dispersion of the SCD quantitatively characterizing its width. The second of these statistical characteristics is extremely significant for the thermodynamics of the melt of a heteropolymer specimen, being in a simple way $\Delta H_{\text{mix}} = RT\chi\sigma_l^2$ connected with the specific enthalpy of mixing ΔH_{mix} per mole of monomeric units. Here T is the absolute temperature, R represents the gas constant, whereas χ denotes the Flory χ-parameter whose values are available from the literature for many pairs of monomeric units (see, for example, [7]).

2.2 Configurational Statistics of Heteropolymers

Statistical characteristics of the chemical structure of a heteropolymer, which pertain to the second type, describe the pattern of arrangement of units along macromolecules. The best known among such characteristics are fractions $P\{U_k\}$ of directed sequences $\{U_k\}$ incorporating k monomeric units. The simplest of them are the dyads $\{U_2\}$, the complete set of which for a binary copolymer is composed of four pairs $\overline{M_1M_1}, \overline{M_1M_2}, \overline{M_2M_1}, \overline{M_2M_2}$. Their calculation turns out to be rather useful for two reasons.

Firstly, some important characteristics of the performance properties of copolymers can be expressed through the fractions of these sequences. For instance, dealing with the thermostability of copolymers, they normally resort to the semi-empirical formulas relating the glass transition temperature, T_g, with fractions of dyads in macromolecules. The simplest among such formulas reads

$$T_g = T_{11}P\{\overline{M_1M_1}\} + T_{22}P\{\overline{M_2M_2}\} + T_{12}\left(P\{\overline{M_1M_2}\} + P\{\overline{M_2M_1}\}\right) , \tag{3}$$

where T_{11} and T_{22} are the glass transition temperature of a homopolymer consisting of units $\overline{\mathrm{M}}_1$ and $\overline{\mathrm{M}}_2$, respectively, and T_{12} is T_g of a heteropolymer in molecules of which these units regularly alternate. Values of T_{12} for a number of copolymers are reported in the literature [8, 9]. Secondly, fractions $P\{U_k\}$ can be determined for $k = 2, 3, 4$ with high accuracy by means of a spectroscopic technique [10, 11]. A comparison of such experimental findings with theoretical results achieved in the framework of the kinetic model chosen permits making definite conclusions about its adequacy for the description of a process for obtaining heteropolymers.

Characterization of the pattern of arrangement of units in chains of heteropolymers in terms of the probabilities of sequences $\{U_k\}$ is of considerable use when handling macromolecules in which the average length of blocks of single-type units is small enough. Just such heteropolymers are formed when they are obtained in traditional ways. More valuable information on the chemical structure of heteropolymers comprising long blocks is provided by the function of distribution of $\alpha = 1, 2$-type blocks for lengths

$$f_\alpha^{\mathrm{bl}}(n) = P\{\overline{\mathrm{M}}_\beta \overline{\mathrm{M}}_\alpha^n \overline{\mathrm{M}}_\beta\}/P\{\overline{\mathrm{M}}_\beta \overline{\mathrm{M}}_\alpha\} \qquad (\alpha \neq \beta)\,. \tag{4}$$

This function is equal to the ratio of the probability of a cluster with length n to the probability of directed dyad $\{\overline{\mathrm{M}}_\beta \overline{\mathrm{M}}_\alpha\}$.

One more quantitative way to characterize the chemical structure of heteropolymers is based on the consideration of chemical correlation functions [1, 3]. The simplest of such chemical correlators is the two-point correlator $Y_{\alpha,\beta}(n)$ which equals the joint probability for two monomeric units divided in a polymer chain by an arbitrary directed sequence $\{U_n\}$ containing n units to be of types α and β

$$Y_{\alpha\beta}(n) = \sum_{\{U_n\}} P\left\{\overline{\mathrm{M}}_\alpha U_n \overline{\mathrm{M}}_\beta\right\}\,. \tag{5}$$

This correlator plays a key role in the thermodynamics of solutions and blends of heteropolymers, since its generating function

$$W_{\alpha\beta}(x) = \sum_{n=0}^{\infty} Y_{\alpha\beta}(n)\, x^{n+1} \tag{6}$$

enters into the expression which describes the angular dependence of the intensity of light or neutron scattering [12, 13]. Looking at the expression for the three-point chemical correlator

$$Y_{\alpha\beta\gamma}(n_1, n_2) = \sum_{\{U_{n_1}\}} \sum_{\{V_{n_2}\}} P\left\{\overline{\mathrm{M}}_\alpha U_{n_1} \overline{\mathrm{M}}_\beta V_{n_2} \overline{\mathrm{M}}_\gamma\right\} \tag{7}$$

it is easy to figure out how it can be extended to the m-point correlator with arbitrary m. Determination of the chemical correlators constitutes one of the major challenges of the statistical chemistry of polymers, because through their generating functions the vertices of the Landau free energy expansion are explicitly expressed as functions of wave vectors [5, 14]. Given these vertices, the standard formalism of the Weak Segregation Theory permits constructing the phase diagram of the melt of any heteropolymer in the vicinity of its critical point [5, 15].

2.3 Markovian Copolymers

The above-mentioned statistical characteristics of the chemical structure of heteropolymers are easy to calculate, provided they are Markovian. Performing these calculations, one may neglect finiteness of macromolecules equating to zero elements $\nu_{\alpha 0}$ of transition matrix $\boldsymbol{Q}$. Under such an approach vector $\boldsymbol{X}$ of a copolymer composition whose components are $X_1 \equiv P(\overline{\mathrm{M}}_1)$ and $X_2 \equiv P(\overline{\mathrm{M}}_2)$ coincides with stationary vector $\boldsymbol{\pi}$ of matrix $\boldsymbol{Q}$. The latter is, by definition, the left eigenvector of this matrix corresponding to its largest eigenvalue λ_1, which equals unity. Components of the stationary vector

$$\pi_1 = \frac{\nu_{21}}{\nu_{12} + \nu_{21}} \qquad \pi_2 = \frac{\nu_{12}}{\nu_{12} + \nu_{21}} \tag{8}$$

can be found from the solution of linear algebraic equations

$$\sum_{\alpha=1}^{2} \pi_\alpha \nu_{\alpha\beta} = \pi_\beta \,, \qquad \sum_{\alpha=1}^{2} \pi_\alpha = 1 \,, \qquad (\beta = 1, 2) \,. \tag{9}$$

The expression for the two-point chemical correlator (Eq. 5) for a Markovian copolymer looks as follows

$$Y_{\alpha\beta}(n) = \pi_\alpha (\boldsymbol{Q}^{n+1})_{\alpha\beta} = \pi_\alpha \sum_{\nu=1}^{2} \lambda_\nu^{n+1} \psi_\alpha^{+\nu} \psi_\beta^{\nu} \,, \tag{10}$$

where $\boldsymbol{\psi}^\nu$ and $\boldsymbol{\psi}^{+\nu}$ are respectively its right and left eigenvectors corresponding to the eigenvalue λ_ν. Expressing them through the elements of matrix $\boldsymbol{Q}$ and components of its stationary vector $\boldsymbol{\pi} = \boldsymbol{X}$ one can readily arrive at the following relationship for two-point chemical correlator

$$Y_{\alpha\beta}(n) = X_\alpha X_\beta + X_1 X_2 (-1)^{\alpha+\beta} (1 - \varepsilon)^{n+1} \,, \tag{11}$$

where quantity $\varepsilon = \nu_{12} + \nu_{21}$ (equal to the sum of reciprocal lengths of blocks of monomeric units) is small for multiblock copolymers with blocks of large length. It is essential that the decay of chemical correlations in the macromolecules of Markovian copolymers is of exponential character.

When considering the distribution for length of the blocks in the copolymer molecules one can proceed from the general definition (Eq. 4). For

Markovian copolymers numerator and denominator in this expression can be reduced to the well-known form

$$P\left\{\overline{\mathrm{M}}_\beta\overline{\mathrm{M}}_\alpha^n\overline{\mathrm{M}}_\beta\right\} = \pi_\beta\nu_{\beta\alpha}\nu_{\alpha\alpha}^{n-1}\nu_{\alpha\beta}\,, \tag{12}$$
$$P\left\{\overline{\mathrm{M}}_\beta\overline{\mathrm{M}}_\alpha\right\} = \pi_\beta\nu_{\beta\alpha}\,, \qquad (\alpha \neq \beta)$$

showing exponential decay with the growth of n. However, exponentials in formula (Eq. 12) for units $\overline{\mathrm{M}}_1$ and $\overline{\mathrm{M}}_2$ being distinct from each other differ from the chemical correlator exponential (Eq. 11) as well.

When considering the composition inhomogeneity of Markovian copolymers, the finiteness of the chemical size of macromolecules cannot be ignored, because fractional composition distribution $\mathrm{W}(l|\zeta)$ in the limit $l \to \infty$ turns out to be equal to the Dirac delta function $\delta(\zeta - X)$. For macromolecules of finite size $l \gg 1$ the function $\mathrm{W}(l|\zeta)$ is the Gaussian distribution whose center and dispersion (Eq. 2) are described by relationships (Eq. 8) and the following one

$$\sigma_l^2 = \frac{X_1X_2}{l}\left(\frac{2}{\varepsilon} - 1\right)\,. \tag{13}$$

Thus, as can be inferred from the foregoing, the calculation of any statistical characteristics of the chemical structure of Markovian copolymers is rather easy to perform. The methods of statistical chemistry [1, 3] can reveal the conditions for obtaining a copolymer under which the sequence distribution in macromolecules will be describable by a Markov chain as well as to establish the dependence of elements $\nu_{\alpha\beta}$ of transition matrix $\boldsymbol{Q}$ of this chain on the kinetic and stoichiometric parameters of a reaction system. It has been rigorously proved [1, 3] that Markovian copolymers are formed in such reaction systems where the Flory principle can be applied for the description of macromolecular reactions. According to this fundamental principle, the reactivity of a reactive center in a polymer molecule is believed to be independent of its configuration as well as of the location of this center inside a macromolecule.

A kinetic model based on the Flory principle is referred to as the ideal model. Up to now this model by virtue of its simplicity, has been widely used to treat experimental data and to carry out engineering calculations when designing advanced polymer materials. However, strong experimental evidence for the violation of the Flory principle is currently available from the study of a number of processes of the synthesis and chemical modification of polymers. Possible reasons for such a violation may be connected with either chemical or physical factors. The first has been scrutinized both theoretically and experimentally, but this is not the case for the second among which are thermodynamic and diffusion factors. In this review we by no means pretend to cover all theoretical works in which these factors have been taken into account at the stage of formulating physicochemical models of the process of synthesis of a heteropolymer. Our task is just to exemplify by our recent publications the importance of the conformational analysis of polymer chains participating in macromolecular reactions.

3 Polymeranalogous Reactions in Dilute Solution

3.1 Main Regularities

In the course of this reaction, the monomeric units constituting a macromolecule change their type due to the entry into the chemical reaction of their functional groups with some low-molecular reagent. If in the beginning there was a homopolymer with units $\overline{M}_1$, each containing one functional group A, then at complete transformation of these groups A $\rightarrow$ B the polymeranalogous reaction (PAR) will result in the formation as its final product of a homopolymer with units $\overline{M}_2$, each containing one group B. However, the products of any incomplete transformation of groups A into B will be heteropolymer chains consisting of units of two types, $\overline{M}_1$ and $\overline{M}_2$. Chemical length l of every such macromolecule remains unaltered in the course of PAR, whereas its composition and chemical structure (i.e., configuration) undergo change.

When the chemical modification of macromolecules takes place in a dilute solution, each may be considered as an isolated "microreactor" into which molecules of low-molecular reagent Z penetrate from outside. Depending on the ratio between rate V_{diff} of its diffusion in the microreactor and rate V_{chem} of its disappearance caused by the interaction with functional groups A, two limiting macrokinetic regimes of a polymeranalogous reaction are distinguished. In the case $V_{\text{diff}} \gg V_{\text{chem}}$, this reaction is kinetically controlled, whereas at $V_{\text{diff}} \ll V_{\text{chem}}$ it is diffusion controlled. If macromolecules in solution have a coil conformation, the first of these regimes is realized. However, PAR in polymer globules may well proceed in the second regime, provided the density of their constituent units is high enough to appreciably retard the reagent Z diffusion. Its rate inside a globule will be especially small when this globule is in a glassy state.

Presently, the quantitative theory of irreversible polymeranalogous reactions proceeding in a kinetically-controlled regime is well along in development [16, 17]. Particularly simple results are achieved in the framework of the ideal model, the only kinetic parameter of which is constant k of the rate of elementary reaction A + Z $\rightarrow$ B. In this model the sequence distribution in macromolecules will be just the same as that in a random copolymer with parameters $P(\overline{M}_1) = X_1 = p$ and $P(\overline{M}_2) = X_2 = 1 - p$ where p is the conversion of functional group A that exponentially depends on time t and initial concentration $\overline{M}_1^0$ of monomeric units of the first type

$$p = \exp\left(-k\overline{M}_1^0 t\right) . \tag{14}$$

Chemical correlations in chains of a random copolymer are entirely absent, as can be seen from formula (Eq. 11) at $\varepsilon = 1$, while the distribution (Eq. 4) of blocks for lengths looks like

$$f_\alpha^{\text{bl}}(n) = X_\alpha^{n-1} X_\beta \qquad (\alpha \neq \beta) \tag{15}$$

The center and dispersion (Eq. 2) of the Gaussian distribution describing the composition inhomogeneity of a random copolymer comprising macromolecules whose length is $l \gg 1$ have the very simple appearance

$$X = 1 - p \qquad \sigma_l^2 = p(1-p)/l . \tag{16}$$

The progress of the theory of polymeranalogous transformations in a dilute solution was basically associated with the extension of the ideal kinetic model connected with the allowance for the dependence of functional group A reactivity on its microenvironment [16]. The latter is largely prescribed by the types of monomeric units neighboring in a macromolecule with the unit to which a given group A is attached. The types of these neighboring units change, once their functional groups enter into the reaction. Along with this chemical factor, the microenvironment of group A is also governed by physical factors, which come into play due to the alteration of local concentrations of monomeric units and low-molecular reagent Z in the vicinity of functional group A. If the rate of their transformation into B-groups is sufficiently slow in comparison with that of reagent Z diffusion, local concentrations $\overline{M}_1, \overline{M}_2$ and Z will evidently have equilibrium values. To find them, Litmanovitch put forward [18, 19] the simplest lattice model originating from the main ideas of the mean field theory. In the framework of this model, the account of the effect of the microenvironment of groups A does not alter the form of the kinetic equations, leading only to the appearance of the dependence of their coefficients on time.

A completely different type of situation occurs under theoretical consideration of a PAR in a dilute solution of dense homopolymer globules, provided the rate of reagent Z diffusion in them is small as compared to that of its chemical reaction with groups A. In the course of such a diffusion-controlled chemical transformation, heteropolymer macromolecules are formed with a rather specific pattern of arrangement of monomeric units. For the first time theoretical consideration of the chemical structure of such heteropolymers was carried out in paper [20]. Its authors found that the stochastic chain describing this structure is not the Markov chain, but obeys the so-called Levy-flight statistics [21]. This conclusion was drawn for a globule whose core consisting of units $\overline{\mathrm{M}}_1$ is separated from a shell comprising units $\overline{\mathrm{M}}_2$ by a sharp boundary. For such a "core-shell" morphology of a globule inside

Fig. 1 Schematic representation of a globule of a binary copolymer with a fuzzy profile of monomeric units

which the concentration profiles of units are of step-wise character, explicit formulas for the distributions of blocks of these units for lengths (Eq. 4) have been derived [20].

In order to estimate the region of this approximation applicability, it is necessary to examine macrokinetics of a polymeranalogous reaction with explicit allowance for the diffusion of a reagent Z into a globule. In this case, the profile of its constituent monomeric units will be fuzzy rather than stepwise (see Fig. 1). This brings up two questions. The first one is how this profile depends on kinetic and diffusion parameters of a reaction system. The second question is concerned with the effect of the profile shape on the statistical characteristics of the chemical structure of the products of a polymeranalogous reaction. A rigorous theory has been developed [22, 23] which enables us to answer these questions. The main concepts of this theory are outlined in the subsequent Sections.

3.2 Macrokinetics of PAR in a Globule

The main macrokinetic problem to be solved for the description of this reaction is finding the evolution of the profile of concentrations $\overline{M}_1$, $\overline{M}_2$ of monomeric units $\overline{\mathrm{M}}_1$, $\overline{\mathrm{M}}_2$ inside a globule with radius R. By virtue of the spherical symmetry of the problem, concentration $\overline{M}_1$ is the same at all points of a globule located at identical distance r from its center. The same condition is apparently met by the concentrations of the second type units $\overline{M}_2 = \overline{M}_1^0 - \overline{M}_1$ and low-molecular reagent Z. Presuming monomeric units to be immobile and the Flory principle to hold, it is possible to write down a simple set of two kinetic equations describing the evolution of concentrations $\overline{M}_1$ and Z in space and time [22]. Along with the elementary reaction constant k and diffusion coefficient D of reagent Z, this macrokinetic model will have as its parameters the equilibrium value Z_e of its concentration at a globule boundary and the concentration $\overline{M}_1^0$ of the first type units in the initial homopolymer globule. Kinetic equations together with the initial and boundary conditions, written down in dimensionless form, read

$$\frac{\partial z}{\partial \tau} = \frac{1}{h^2}\left(\frac{1}{\rho^2}\frac{\partial}{\partial \rho}\rho^2\frac{\partial z}{\partial \rho}\right) - w_1 z \tag{17}$$

$$w_1(\rho,\tau) = \exp\left\{-b\int_0^\tau z(\rho,\theta)\,d\theta\right\}; \qquad w_2 = 1 - w_1$$

$$z(\rho,0) = 0\,, \qquad z(0,\tau) < \infty\,, \qquad z(1,\tau) = 1\,, \tag{18}$$

where the following designations are used

$$z = \frac{Z}{Z_e}\,, \qquad w_1 = \frac{\overline{M}_1}{\overline{M}_1^0}\,, \qquad w_2 = \frac{\overline{M}_2}{\overline{M}_1^0}\,, \tag{19}$$

$$\rho = \frac{r}{R}\,, \qquad \tau = k\overline{M}_1^0 t\,.$$

The solution of mathematical model (Eqs. 17, 18) is controlled exclusively by two dimensionless parameters

$$h = R\sqrt{\frac{k\overline{M}_1^0}{D}} \qquad b = \frac{Z_e}{\overline{M}_1^0} \tag{20}$$

the first of which, i.e., the Thiele modulus, is well-known in the macrokinetics of chemical reactions [24]. The second parameter, b, is of thermodynamic nature, because the value of Z_e can be calculated from the condition of the equality of chemical potential of molecules Z on the boundary of a globule and outside it. Below for the sake of simplicity the consideration will be confined to the systems where the distinction in equilibrium concentration of reagent Z in globules with different proportions of units $\overline{\mathrm{M}}_1$ and $\overline{\mathrm{M}}_2$ is negligible. In this case the quantity Z_e may be thought of as independent of the concentration of units $\overline{\mathrm{M}}_1$, and thus as time independent.

When considering the macrokinetics of PAR described by equations (Eq. 17), it is reasonable to focus on two limiting regimes. The first of these, the kinetically-controlled regime, takes place provided the rate of diffusion of molecules Z appreciably exceeds that of the chemical reaction. In this case, a uniform concentration $Z = Z_e$ should be established all over the globule after time interval $t_d \sim R^2/D$. Subsequently, during the interval $t_1 \sim 1/kZ_e$, which is considerably larger than t_d, the transformation of units $\overline{\mathrm{M}}_1$ into $\overline{\mathrm{M}}_2$

will happen with the same rate all over the globule. Therefore, the concentration of units $\overline{\mathrm{M}}_1$ does not depend on r and will decrease exponentially with time $\overline{M}_1(t) = \overline{M}_1^0 \exp(-kZ_e t)$. This regime is realized if $t_d \ll t_z \sim 1/k\overline{M}_1^0$.

When the reverse inequality $t_d \gg t_z$ holds, the other macrokinetic regime, i.e., a diffusion-controlled one, takes place. For such a regime, the reaction proceeds only within a narrow layer whose thickness is much smaller than the globular radius. In the asymptotic limit of the instantaneous reaction $t_z \to 0$, the profile of the concentration of units $\overline{\mathrm{M}}_1$ is step-wise. Within the core of the globule ($0 < r < R^*$) the concentration $\overline{M}_1(r) = \overline{M}_1^0$, whereas in the external part of the globule ($R^* < r < R$) this concentration is equal to zero. The reaction front $R^*(t)$ separating the core and the shell of the globule moves from its border inward according to the law [25]

$$R^*(t) = R\left(1 - 2\sqrt{qt/t_d}\right), \tag{21}$$

where the dependence of the variable q on the parameter $b = Z_e/\overline{M}_1^0$ is calculated by the equation

$$\sqrt{\pi q}\exp(q)\mathrm{erf}(\sqrt{q}) = b, \text{ where } \mathrm{erf}\xi = \frac{2}{\sqrt{\pi}}\int_0^{\xi} e^{-\eta^2}\mathrm{d}\eta. \tag{22}$$

In the most interesting case, $b \ll 1$, the probability integral, $\mathrm{erf}(\sqrt{q})$, equals $2\sqrt{q/\pi}$, so that $q = b/2$ and thus the time of completion of the reaction will have the order of magnitude t_d/b^2.

The two regimes discussed above describe the reaction in a globule, provided the distinction in scales of characteristic times of diffusion, t_d, and reaction of low-molecular reagent Z, t_z, is large enough. If the difference between t_d and t_z is not so pronounced, the concentration of this reagent as well as that of units $\overline{\mathrm{M}}_1$ and $\overline{\mathrm{M}}_2$ will gradually change outward from the center of the globule. The evolution of the profile of these concentrations is described by the solution of kinetic equations (Eq. 17).

To the asymptotic values $h \to 0$ and $h \to \infty$ of the Thiele modulus, which is one of the parameters of these equations, there correspond two limiting macrokinetic regimes of the reaction discussed above. Some examples of the concentration profiles of the reagents at various values of parameters (Eq. 20) are presented in Fig. 2. This figure shows the transition from one limiting case $h \to 0$ to another $h \to \infty$. The larger the Thiele modulus is, the narrower is the region of p where both concentration of reagents A and Z are appreciably distinct from zero. Figure 2 illustrates how the width of this reaction zone approaches zero as parameter h tends to infinity.

As for the dependence of the profile $w_1(\rho)$ on parameter b, it is not essential for its small values. Authors of paper [22] arrived at this conclusion having superimposed the profiles corresponding to equimolar composition

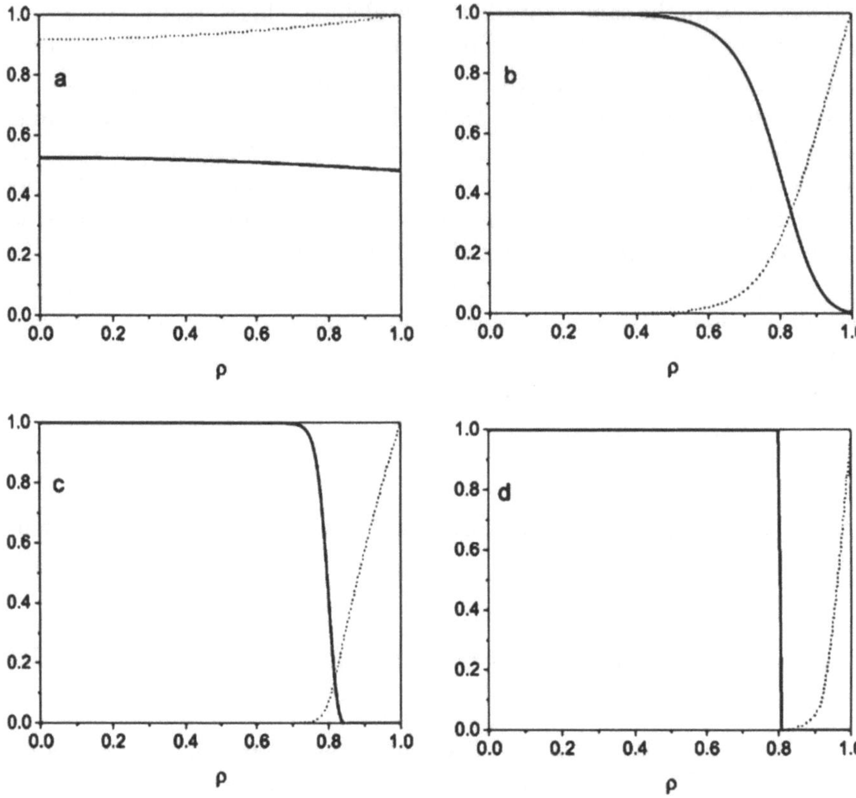

Fig. 2 Dependence of dimensionless concentrations of type 1 monomeric units w_1 (*thick line*) and low-molecular compound z (*thin line*) on reduced distance ρ from the globule center at values of the Thiele modulus (Eq. 20) $h = 1$ (**a**), 15 (**b**), 60 (**c**) and ∞ (**d**). ($X_1 = X_2 = 0.5$, value of parameter b (Eq. 20) equals 0.01)

of a copolymer obtained at the same values of the Thiele modulus, when b is equal to 0.01, 0.05, and 0.1. All curves calculated for parameter h equal to 10, virtually coincided. The same result was obtained for the Thiele modulus equal to 70.

3.3 Statistical Physics of Dense Heteropolymer Globule

The chemical structure of any heteropolymer macromolecule is defined by its configuration $\{\alpha_i\}$ where $\alpha_i = 1, 2$ denotes the type of monomeric unit. A special feature of this structure in the system in hand is that its formation is largely conditioned by the conformational state of macromolecules involved. Hence, under the statistical consideration of these macromolecules it was suggested [22] to differentiate polymer chains apart from configuration $\{\alpha_i\}$ also by their conformation $\{r^i\}$ that is characterized by a set of spatial coordinates $(r^1, \ldots, r^l)$ of all its l units. Authors [22] proceeded from the assumption that the distribution of probabilities $\mathcal{P}_l\{r^i\}$ of macromolecular conformations $\{r^i\}$ remains unaltered during the reaction and is equal to that occurring in a homopolymer globule. An analogous assumption was implicitly employed earlier in papers [20, 26–29] where use was made of a formal procedure of coloring in two colors the core and the shell of the initial globule. In terms of the system under study these colors correspond to monomeric units $\overline{\mathrm{M}}_1$ and $\overline{\mathrm{M}}_2$.

Within the framework of the present model, the molecular-structure distribution of a proteinlike heteropolymer is determined through the expression

$$\mathcal{P}_l\{\alpha_i\} = \int \cdots \int \mathcal{P}_l\{r^i\} \prod_{i=1}^{l} w_{\alpha_i}(r^i)\mathrm{d}r^i . \tag{23}$$

Its derivation implies a succession of two formal procedures. First, it is necessary to color the homopolymer globule units marking every i-th unit by color α_i with the probability $w_{\alpha_i}(r^i)$ which coincides with the ratio of concentration of units $\overline{\mathrm{M}}_{\alpha_i}$ at point r^i to the overall concentration of all units at this point. As a result of such a coloring, the joint distribution for configurations and conformations of proteinlike heteropolymers is obtained. Integration of this distribution over coordinates of all units results in the desired molecular-structure distribution (Eq. 23).

The distribution of probabilities of macromolecular conformations in the globular state is well known [30]

$$\mathcal{P}_l\{r^i\} = Z_l^{-1}\,[\{H\}]\, e(r^1) \prod_{i=2}^{l} Q(r^{i-1}, r^i) . \tag{24}$$

Here the following designations are used

$$e(r^i) \equiv \exp\left[-H(r^i)/T\right] , \qquad Q(r^{i-1}, r^i) \equiv \lambda(r^{i-1} - r^i)e(r^i) , \tag{25}$$

where $H(r^i)$ is a self-consistent field acting on the i-th unit from other units of a macromolecule while $\lambda(r^{i-1} - r^i)$ stands for the probability for neighboring units to be located at points r^{i-1} and r^i. The partition function Z_l is found from the condition that the integral of conformation distribution $\mathcal{P}_l\{r^i\}$ over coordinates of all units should be equal to unity

$$Z_l\,[\{H\}] = \int \cdots \int e(r^1)\mathrm{d}r^i \prod_{i=2}^{l} Q(r^{i-1}, r^i)\mathrm{d}r^i . \tag{26}$$

The partition function (Eq. 26) may be regarded as a functional of external field $\{H\} = H(r^1), \ldots, H(r^l)$, which can be determined from the self-consistence condition [30]. Of utmost importance in finding the chemical correlators

(Eqs. 5, 7) is the generating function of MSD (Eq. 24)

$$G_l\{s^i\} \equiv \sum_{\{\alpha_k\}} P_l\{\alpha_k\} \prod_{k=1}^{l} s^k_{\alpha_k} = Z_l\left[\{H^i_{\mathrm{gh}}\}\right]/Z_l\left[\{H\}\right] \qquad G_l\{1\} = 1\,, \tag{27}$$

where the numerator is the partition function of a homopolymer molecule whose i-th unit is influenced by the external ghost field

$$H^i_{\mathrm{gh}}(r) = H(r) + H^i(r)\,, \qquad H^i(r) = -\,T\ln\sum_\alpha w_\alpha(r)s^i_\alpha\,. \tag{28}$$

This field represents the sum of the self-consistent field of physical interactions of units $H(r)$ and virtual field $H^i(r)$ related to their coloring. The latter vanishes provided all $s^i_\alpha = 1$, because the sum of probabilities $w_\alpha(r)$ is identically equal to unity at any point r.

If number l of units in a globule is large, the density of units can be determined from equations

$$\rho(r) = l\psi(r)\psi^+(r)\,, \qquad \psi(r) = \psi^+(r)e(r) \tag{29}$$

where $\psi(r)$ and $\psi^+(r)$ stand for orthonormalized eigenfunctions of integral operator $\mathbf{Q}$ with kernel $Q(r', r'')$ (Eq. 25) and conjugate operator $\mathbf{Q}^+$ with kernel $Q^+(r', r'') = Q(r'', r')$, corresponding to the largest eigenvalue Λ

$$\int \psi(r')Q(r', r'')\mathrm{d}r' = \Lambda\psi(r'')\,, \qquad \int \psi^+(r')Q^+(r', r'')\mathrm{d}r' = \Lambda\psi^+(r'')\,. \tag{30}$$

Of prime importance in the theory of polymer globules is the propagator $Q^{(n)}(r', r'')$ representing the kernel $Q(r', r'')$ of the operator $\mathbf{Q}$ iterated n times. Let Λ_ν ($\nu = 0, 1, \ldots$) be the eigenvalues of this operator in order of their decreasing, while $\psi^\nu(r)$ and $\psi^{\nu+}(r)$ stand for corresponding eigenfunctions of operators $\mathbf{Q}$ and $\mathbf{Q}^+$. The propagator can be written as follows

$$Q^{(n)}(r', r'') = \sum_\nu \Lambda^n_\nu \psi^{\nu+}(r')\psi^\nu(r'') = \Lambda^n\psi^+(r')\psi(r'')\,, \tag{31}$$

where the second equality holds only in the asymptotic limit $n \to \infty$ when all items in the sum are exponentially small in comparison with the first one. Index $\nu = 0$ in expression (Eq. 31) is dropped for the sake of simplicity. Using this expression in the case $l \gg 1$ it is easy to get the asymptotic formula for the partition function

$$Z_l = \iint \mathrm{d}r^1\mathrm{d}r^l e(r^1)Q^{(l-1)}(r^1, r^l) = C^2\Lambda^{l-1} \tag{32}$$

where constant C stands for the integral of function $\psi(r)$. Equation 29 can be derived in a standard manner

$$\rho(r) = \frac{\delta F_l[H]}{\delta H(r)} = -\frac{T}{Z_l}\frac{\delta Z_l[H]}{\delta H(r)} \tag{33}$$
$$= \frac{1}{Z_l}\sum_{i=1}^{l}\iint \mathrm{d}r^1 \mathrm{d}r^l Q^{+(i-1)}(r^1,r)e(r)Q^{(l-1)}(r,r^l) = l\psi(r)\psi^+(r)$$

since the density is a functional derivative of free energy $F_l[H] = -T\ln Z_l[H]$ with respect to the external field. The last equality is valid at sufficiently large values of l when the contribution into the sum (Eq. 33) from units situated close to the ends of the macromolecule is negligible. In this case, one can use the asymptotic expression (Eq. 31) for the propagators of the integral (Eq. 33).

In the theory of the polymer globule, it was shown [30], that for macromolecules with a large number of units l this variable may be considered as continuous. In such a continuous model the integral operator $\mathbf{Q}$ is replaced by the differential one $\mathbf{Q}_{\text{diff}} = e(r)[1 + (a^2/6)\Delta]$, where Δ is the Laplacian. The spectrum of the operator $\mathbf{Q}_{\text{diff}}$ has an especially simple form provided the distribution of the density of units in a globule is approximated by the step-function $\rho(r) = \rho^0\eta(R-r)$, where the value of ρ^0 can be obtained from the condition of vanishing of the osmotic pressure inside a globule. In this volume approximation eigenvalues of operator $\mathbf{Q}_{\text{diff}}$ look as follows

$$\Lambda_\nu = \Lambda(1-\varepsilon_\nu), \qquad \varepsilon_\nu = \xi_\nu^2\varepsilon \tag{34}$$
$$\varepsilon = \frac{a^2}{6R^2} = \frac{1}{6}\left(\frac{4\pi\Phi_g}{3l}\right)^{2/3}, \qquad \Phi_g = \frac{4\pi R^3}{3}\rho_0,$$

where ξ_ν is the ν-th root ($\nu = 0, 1, \ldots$) of equation $\xi_\nu = \tan\xi_\nu$ and Φ_g stands for the volume fraction of monomeric units in a dense globule of radius R. Explicit expressions for the orthonormalized eigenfunctions of the operator $\mathbf{Q}_{\text{diff}}$ have a rather simple appearance [22].

3.4 Chemical Structure of Proteinlike Macromolecules

3.4.1 Chemical Correlators

Since the molecular-structure distribution (Eq. 23) contains exhaustive information on the chemical structure of a heteropolymer, knowledge of generating the function of MSD (Eq. 27) suffices for finding any statistical characteristic of this structure. When obtaining the formula for the generating function of chemical correlators, it is natural to neglect the end-effects as was done when deriving the expression for the density of units in a globule

(Eq. 33). Differentiating the function (Eq. 27) and setting the components of all vectors s^i equal to unity, we get

$$\frac{\partial G_l}{\partial s^i_\alpha} = \int \mathrm{d}r^i \psi(r^i) w_\alpha(r^i) \psi^+(r^i) = \frac{1}{l} \int \mathrm{d}r \rho(r) w_\alpha(r) \tag{35}$$

$$\frac{\partial^2 G_l}{\partial s^i_\alpha \partial s^j_\beta} = \iint \mathrm{d}r^i \mathrm{d}r^j \psi(r^i) w_\alpha(r^i) Q_R^{(j-i)}(r^i, r^j) w_\beta(r^j) \psi^+(r^j)\,. \tag{36}$$

In the right-hand side of the expression (Eq. 36) we have $(j-i)$ times iterated kernel of reduced operator $\mathbf{Q}_R$ whose kernel and eigenvalues read

$$Q_R(r', r'') = Q(r', r'')\Lambda^{-1}\,, \qquad \Lambda^R_\nu = \Lambda_\nu \Lambda^{-1} = 1 - \varepsilon_\nu\,, \tag{37}$$

where ε_ν is defined above (Eq. 34).

Expression (Eq. 35) has the sense of the probability P^i_α for a unit located at i-th position to be of type α. Quantity P^i_α according to (Eq. 35) turns out to be independent of i and equal to X_α. It means that the stochastic process R of conventional movement along macromolecules, describing their chemical structure, is stationary [17]. This statement does not concern the description of end-fragments of a polymer chain at the scale corresponding to the decay of chemical correlators. In the limit $l \to \infty$ expressions (Eq. 35) and (Eq. 36) are asymptotically exact expressions for the composition of a copolymer and its two-point chemical correlator $Y_{\alpha\beta}(n)$. The latter can be expressed through eigenvalues and eigenfunctions of operator $\mathbf{Q}$, if the recourse is made to the formulas (Eqs. 31, 36 and 37)

$$Y_{\alpha\beta}(n) = \sum_{\nu=0} (\Lambda^R_\nu)^{n+1} \theta^{0\nu}_\alpha \theta^{\nu 0}_\beta = Y_{\beta\alpha}(n)\,, \tag{38}$$

where the following designation is used

$$\theta^{\nu\mu}_\alpha = \int \mathrm{d}r \psi^\nu(r) w_\alpha(r) \psi^{\mu +}(r) = \theta^{\mu\nu}_\alpha\,. \tag{39}$$

Eigenvalues of the operator $\mathbf{Q}_R$ are real while the largest of them, Λ^R_0, equals unity by definition. As a result, in the limit $n \to \infty$ all items in the sum (Eq. 38), excluding the first one, $\theta^{00}_\alpha \theta^{00}_\beta = X_\alpha X_\beta$, will vanish. In this case, chemical correlators will decay exponentially along the chain on the scale $n^* \sim -1/\ln \Lambda^R_1$. At values $n < n^*$ the law of the decay of these correlators differs, however, from the exponential one even for binary copolymers. This obviously testifies to non-Markovian statistics of the sequence distribution in molecules (see expression Eq. 11). The closer is Λ^R_1 to unity, the greater are the values of n^*. The situation when $n^* \gg 1$ corresponds to proteinlike copolymers.

Quite special is the consideration of the kinetically-controlled regime of the reaction in a globule. For this limiting macrokinetic regime probability

w_α is independent of r and coincides with X_α

$$X_\alpha = \int dr\psi(r)w_\alpha(r)\psi^+(r)\,. \tag{40}$$

As to quantities $\theta_\alpha^{\nu\mu}$ (Eq. 39), they are equal to $X_\alpha\delta_{\nu\mu}$, where $\delta_{\nu\mu}$ is the Kronecker symbol. In view of this, only the first item in the sum (Eq. 38), $X_\alpha X_\beta$, is distinct from zero. This corresponds to the absence of chemical correlations along the macromolecule on all scales. In other words, the distribution of units for such copolymers is Bernoullian [2].

It is easy to derive expressions for other chemical correlators as well. For instance, the three-point correlator (Eq. 7) can be determined by the formula

$$\frac{\partial^3 G_l\{s^k\}}{\partial s_\alpha^i \partial s_\beta^j \partial s_\gamma^k} = \sum_\nu \sum_\mu (\Lambda_\nu^R)^{j-i}(\Lambda_\mu^R)^{k-j}\theta_\alpha^{0\nu}\theta_\beta^{\nu\mu}\theta_\gamma^{\mu 0}\,, \tag{41}$$

where quantities $\theta_\alpha^{\nu\mu}$ were defined above (Eq. 39). Since the stochastic process $\mathcal{R}$ of conventional movement along a macromolecule is a stationary one, it is obvious that its three-point correlator $Y_{\alpha\beta\gamma}(n_1, n_2)$ depends only on distances $n_1 + 1 = j - i$ and $n_2 + 1 = k - j$ between units along a macromolecule but not on their positions i, j, k within the polymer chain.

The expressions for generating functions of two- and three-point correlators (Eq. 38), (Eq. 41) read

$$W_{\alpha\beta}(x) = g_l(x)X_\alpha X_\beta + \sigma_{\alpha\beta}(x)\,, \text{ where } \sigma_{\alpha\beta}(x) = \sum_{\nu=1} g_l(\Lambda_\nu^R x)\theta_\alpha^{0\nu}\theta_\beta^{\nu 0} \tag{42}$$

$$W_{\alpha\beta\gamma}(x_1, x_2) = \sum_{\nu=0}\sum_{\mu=0} g_l(\Lambda_\nu^R x_1)g_l(\Lambda_\mu^R x_2)\theta_\alpha^{0\nu}\theta_\beta^{\nu\mu}\theta_\gamma^{\mu 0}\,. \tag{43}$$

Here the following notation is adopted

$$g_l(u) = \frac{u}{l(1-u)}\left(l - \frac{1-u^l}{1-u}\right). \tag{44}$$

The quantities involved in expressions (Eqs. 42 and 43) are defined by formulas (Eqs. 39 and 44) and are controlled apart from the distributions of concentration of different monomeric units in a globule only by eigenvalues and eigenfunctions of the integral operator $\mathbf{Q}$ with kernel (Eq. 25).

Essentially, expressions (Eq. 34) have a physical sense only for values of quantities ε_ν which are appreciably less than unity when the eigenvalues of operators $\mathbf{Q}_{\text{diff}}$ and $\mathbf{Q}$ coincide. The number of such values of ε_ν will be the larger, the longer is the macromolecule. Hence, in the asymptotic limit $l \to \infty$, expression (Eq. 38) for the two-point chemical correlator is reduced to the following form

$$Y_{\alpha\beta}(n) = X_\alpha X_\beta + (-1)^{\alpha+\beta}\sum_{\nu=1}\exp(-\varepsilon_\nu n)(\theta^{0\nu})^2 \tag{45}$$

comprising quantities ε_ν and $\theta^{0\nu} \equiv \theta_1^{0\nu} = -\,\theta_2^{0\nu}$. The latter are defined by expression

$$\theta^{0\nu} = \sqrt{6}C_\nu \int_0^1 \rho \sin(\xi_\nu \rho) w_1(\rho) \mathrm{d}\rho\,, \text{ where } C_\nu = \frac{\sqrt{1+\xi_\nu^2}}{\xi_\nu}\,. \tag{46}$$

Calculating other chemical correlators along with $\theta_\alpha^{00} = X_\alpha$ and $\theta_\alpha^{0\nu}$ (Eq. 46), one should additionally know quantities $\theta_\alpha^{\nu\mu}$ at nonzero values of both superscripts

$$\theta_\alpha^{\nu\mu} = 2C_\nu C_\mu \int_0^1 \sin(\xi_\nu \rho) \sin(\xi_\mu \rho) w_\alpha(\rho) \mathrm{d}\rho\,. \tag{47}$$

The substitution of expressions (Eqs. 46 and 47) and Λ_ν^R into formulas (Eqs. 42 and 43) permits finding the generating function of two- and three-point chemical correlators. Extension of these results to an arbitrary m-point correlator is obvious.

In order to analyze the rate of the decay of chemical correlations in macromolecules of proteinlike copolymers, it is necessary to calculate the dependence on n of the irreducible correlator $\omega_{\alpha\beta}(n) = Y_{\alpha\beta}(n) - X_\alpha X_\beta$ where the reducible correlator $Y_{\alpha\beta}(n)$ is obtained by formula (Eq. 45). Correlator $\omega_{\alpha\beta}(n)$ is calculated as the sum of series (Eq. 45), each item of which is the product of two cofactors. The first of them is controlled just by the number of monomeric units l in a globule, as well as by their volume fraction Φ, whereas the second cofactor is characterized only by the profile $w_1(\rho)$.

The sequence distribution in the copolymers under consideration can not be described by any Markov chain, that may be easily verified by comparing the expressions (Eqs. 11 and 45) for the chemical correlators. A qualitative distinction in chemical structure of the products of the PAR in hand from the Markovian copolymers is that the former are characterized by an infinite set of the chemical correlation lengths, whereas the latter ones are characterized by a single such length. This property inherent in the above copolymers signals that a stochastic chain describing their chemical structure belongs to the stochastic processes known as Levy-flights [21]. This fact was first established in paper [20] by an analytical approach as well as by computer simulation in the case of the "core-shell" profile of the reacted groups inside the dense globule.

Among all correlation lengths entering in expression (Eq. 45) centrally important is the largest one, $n^* \equiv \varepsilon_1^{-1} = \xi_1^{-2}\varepsilon^{-1}$, because just on scale $n \gg n^*$ the chemical correlations in macromolecules of the proteinlike heteropolymers decay. On this scale, the chemical structure of such heteropolymers resembles that exhibited by the Markovian copolymers (see Fig. 3). The only distinction is that for the former the quantity n^* rises with the length l of a macro-

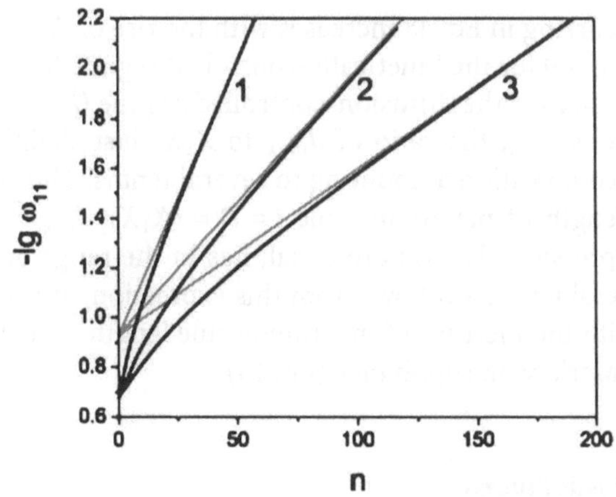

Fig. 3 Dependence of the decimal logarithm of irreducible correlator $\omega_{11} = Y_{11}(n) - X_1^2$ on distance n along polymer chain for macromolecules of equimolar composition at the values of parameters (Eq. 20) $h = 70$, $b = 0.01$. *Curves* are presented for macromolecules whose length is $l = 2 \times 10^3$ (1); 6×10^3 (2); 12×10^3 (3). *Thin lines* represent asymptotic expressions at $n \gg n^*$. Volume fraction of monomeric units in the globule is $\Phi_g = 0.9$

molecule as $n^* \sim l^{2/3}$, whereas for the latter the value of n^* is not controlled by l. Since the average length of blocks constituting copolymers whose composition X is close to 0.5 has the same order of magnitude as n^*, a rough estimate shows that heteropolymer macromolecules of size $l \sim 10^3$ obtained by PAR in a dense globule will largely be composed of 10^1 blocks with length 10^2. This estimate does not contradict the exact solution [20, 23] of the problem of finding the distributions (Eq. 4) of block lengths for the proteinlike heteropolymers.

3.4.2
Composition Distribution

One of the prime objectives of the statistical chemistry of these polymers is establishing the dependence of their composition inhomogeneity on a macromolecule length l and on the reaction system parameters. A quantitative measure of this inhomogeneity is the dispersion (Eq. 2) of the composition distribution

$$\sigma_l^2 = \frac{X_1 X_2}{l} + \frac{d}{l^{1/3}}, \text{ where } d = 12\left(\frac{3}{4\pi\Phi_g}\right)^{2/3} \sum_{\nu=1} \left(\frac{\theta^{0\nu}}{\xi_\nu}\right)^2, \tag{48}$$

which represents a sum of two items. The first of them, equal to σ_l^2 of a random copolymer, is of importance only for moderate values of l. Obviously, for macromolecules whose length is large enough, the second item will dominate.

Parameter d occurring in Eq. 48 increases with the rise of the Thiele modulus h (Eq. 20) from $d = 0$ for the kinetically-controlled regime ($h = 0$) up to some maximal value d_{max} for the diffusion-controlled regime ($h \to \infty$). According to the calculations [23], the ratio of d_{max} to $X_1 X_2$ just slightly depends on the copolymer composition amounting to several tenths. This means that the characteristic length of macromolecule $l = l^* = (X_1 X_2 / d_{max})^{3/2}$, when both items in the expression (Eq. 47) are equal, lies in the range between several unities and several tens. As follows from this expression, at $l \gg l^*$ dispersion σ_l^2 decreases with the increase of macromolecule length l much more slowly than that for a Markovian copolymer (Eq. 13).

3.4.3
Markovian Stochastic Process

For many synthetic copolymers, it becomes possible to calculate all desired statistical characteristics of their primary structure, provided the sequence is described by a Markov chain. Although stochastic process $\mathcal{R}$ in the case of proteinlike copolymers is not a Markov chain, an exhaustive statistic description of their chemical structure can be performed by means of an auxiliary stochastic process $\mathcal{R}_{\text{lb}}$ whose states correspond to labeled monomeric units. As a label for unit $\overline{\text{M}}_\alpha$, it was suggested [23] to use its distance r from the center of the globule. The state of this stationary stochastic process $\mathcal{R}_{\text{lb}}$ is a pair of numbers, (α, r), the first of which belongs to a discrete set while the second one corresponds to a continuous set. Stochastic process $\mathcal{R}_{\text{lb}}$ is remarkable for being stationary and Markovian. The probability of the transition from state (α, r') to state (β, r'') for the process of conventional movement along a heteropolymer macromolecule is described by the matrix-function of transition intensities

$$K_{\alpha\beta}(r', r'') = \left[\psi^+(r')\right]^{-1} Q_R r', r'') w_\beta(r'') \psi^+(r'') . \qquad (49)$$

The role of the latter for proteinlike copolymers is analogous to that played by the transition matrix for Markovian copolymers. This means that $K_{\alpha\beta}(r', r'')$ represents the kernel of integro-matrix operator $\widehat{K}$ whose maximum eigenvalue equals to unity, while corresponding left and right eigenfunctions have the components equal to $\Pi_\alpha(r)$ and 1. As an analog of algebraic equations (Eq. 9), the equations

$$\sum_{\alpha=1}^{2} \int \mathrm{d}r' \Pi_\alpha(r') K_{\alpha\beta}(r', r) = \Pi_\beta(r) \qquad (\beta = 1, 2) \qquad (50)$$

hold, whose solution

$$\Pi_\alpha(r) = \psi(r) w_\alpha(r) \psi^+(r) \qquad (\alpha = 1, 2) \qquad (51)$$

has the meaning of the density of probability for an α-th type monomeric unit to be formed at point r. In order to find the composition of a proteinlike copolymer, it is necessary to erase the label r in the expression for the stationary vector of operator $\widehat{K}$, i.e., to carry out the integration over all values of variable r. This procedure immediately results in formula (Eq. 40).

A similar algorithm that allows finding the statistical characteristics of the chemical structure of non-Markovian heteropolymers is known to be rather efficient for the solution of various problems of statistical chemistry [1]. To realize this algorithm, one should assign a corresponding label to each monomeric unit, find first the expression for each desired statistical characteristic in macromolecules with labeled units and erase eventually all labels.

The potentiality of this algorithm is easy to exemplify by the derivation of the expression (Eq. 38) for the two-point chemical correlator (Eq. 5) of proteinlike copolymers. For the macromolecules with labeled units such a correlator can be obtained proceeding from apparent formula

$$\begin{aligned} Y^{\mathrm{lb}}_{\alpha\beta}(n;r',r'') &= \Pi_\alpha(r')K^{(n+1)}_{\alpha\beta}(r',r'') \\ &= \psi(r')w_\alpha(r')Q^{(n+1)}_R(r',r'')w_\beta(r'')\psi^+(r'') \end{aligned} \tag{52}$$

resembling expression (Eq. 10) which describes Markovian copolymers. Using the spectral expansion of the operator $\widehat{Q}_R$ and erasing the labels on units, one arrives at expression (Eq. 38).

Exploiting the Markovian property of random process $\mathcal{R}_{\mathrm{lb}}$, it is possible to derive in a standard way the expression for an arbitrary chemical correlator. In particular, for the three-point correlator the expression

$$Y^{\mathrm{lb}}_{\alpha\beta\gamma}(n_1,n_2;r',r'',r''') = \Pi_\alpha(r')K^{(n_1+1)}_{\alpha\beta}(r',r'')K^{(n_2+1)}_{\beta\gamma}(r'',r''') \tag{53}$$

defines, along with Eqs. 49 and 51, this function in the ensemble of macromolecules with labeled units. Subsequent label erasing and the recourse to spectral expansion of the operator Q_R leads to the expression (Eq. 41). The algorithm of writing down the formula for any chemical correlator is quite evident. Its employment makes straightforward the procedure of the derivation of the expression

$$\begin{aligned} &W_{\alpha_1\alpha_2\cdots\alpha_m}(x_1,x_2,\ldots,x_{m-1}) \\ &= \sum_{\nu_1}\sum_{\nu_2}\cdots\sum_{\nu_{m-1}} \theta^{0\nu_1}_{\alpha_1} g_l\left(\Lambda^R_{\nu_1}x_1\right)\theta^{\nu_1\nu_2}_{\alpha_2} g_l\left(\Lambda^R_{\nu_2}x_2\right)\cdots g_l\left(\Lambda^R_{\nu_{m-1}}x_{m-1}\right)\theta^{\nu_{m-1}0}_{\alpha_m} \end{aligned} \tag{54}$$

for the generating function of a m-point correlator.

3.5
Scattering and Thermodynamic Behavior of Proteinlike Heteropolymers

3.5.1
Amplitude of Scattering

In order to calculate in the framework of Random Phase Approximation the intensity $I(\theta)$ of scattering at angle θ of the incident radiation with wavelength λ recourse should be made to the formula [31]

$$I(\theta) = \rho^0 \frac{(a_1 - a_2)^2}{\mathrm{D}(Q)}, \text{ where } Q = \frac{a^2}{6}\left(\frac{4\pi}{\lambda}\sin\frac{\theta}{2}\right)^2 . \tag{55}$$

Here a_1 and a_2 stand for the length of scattering of monomeric units $\overline{\mathrm{M}}_1$ and $\overline{\mathrm{M}}_2$, while function $\mathrm{D}(Q)$ is defined as follows

$$\mathrm{D}(Q) = H(Q) - 2\chi, \text{ where } H(Q) = \frac{\widetilde{X}_{11} + \widetilde{X}_{22} + 2\widetilde{X}_{12}}{\widetilde{X}_{11}\widetilde{X}_{22} - \widetilde{X}_{12}^2} \tag{56}$$

$$\widetilde{X}_{\alpha\beta}(Q) = X_\alpha\delta_{\alpha\beta} + W_{\alpha\beta}(x) + W_{\beta\alpha}(x), \qquad x = \exp(-Q). \tag{57}$$

Since the theory under examination works exclusively on scales essentially exceeding size a of a monomeric unit, the function $\mathrm{D}(Q)$ has a physical meaning only at $Q \ll 1$. In this region of variable $\boldsymbol{Q}$ the dependence of elements of matrix $\widetilde{\mathbf{X}}$ (Eq. 57) on the wave vector can be calculated via the relationship

$$\widetilde{X}_{\alpha\beta}(Q) = X_\alpha\delta_{\alpha\beta} + lX_\alpha X_\beta g_{\mathrm{D}}(lQ) + l(-1)^{\alpha+\beta}2\sum_{\nu=1}(Q+\varepsilon_\nu)^{-1}\theta^{0\nu}\theta^{\nu 0}. \tag{58}$$

The simplification, which enables reducing expression (Eq. 42) into (Eq. 58) remains in force in considering the generating function of the arbitrary chemical correlator. This means that in order to use expression (Eq. 54) for the calculation of the dependence of the vertices of the Landau free energy expansion on wave vectors at region $Q_s \ll 1$ $(s = 1, \ldots, m-1)$, one should replace $g_l(\Lambda^R_{\nu_s}x_s)$ by $(Q_s + \varepsilon_{\nu_s})^{-1}$.

For Markovian block copolymers the formula is valid analogous to formula (Eq. 58), however factor $(Q+\varepsilon)^{-1}$ stands instead of the sum over index ν. Thus, in the region of angles θ, at which the inequality $Q \ll \varepsilon_1$ holds, the angular dependence of the amplitude of scattering $I(\theta)$ of a melt of proteinlike heteropolymers turns out to be identical to that of Markovian heteropolymers with $\varepsilon = \varepsilon_1$. Essentially, this coincidence disappears for the values of angle θ at which inequality $Q \ll \varepsilon_1$ is not valid.

3.5.2
Spinodal Equation

The spinodal represents a hypersurface within the space of external parameters where the homogeneous state of an equilibrium system becomes thermodynamically absolutely unstable. The loss of this stability can occur with respect to the density fluctuations with wave vector either equal to zero or distinct from it. These two possibilities correspond, respectively, to trivial and nontrivial branches of a spinodal. The Lifshitz points are located on the hyperline common for both branches.

Bearing in mind that the amplitude of scattering I becomes infinite on the spinodal, it is possible to write down simple conditions for its trivial (a) and nontrivial (b) branches

$$(a)\ \mathrm{D}(0)=0\,,\quad (b)\ \mathrm{D}(Q)=0\,,\quad \mathrm{D}'(Q)=0\,,\quad \mathrm{D}''(Q)>0\,, \tag{59}$$

where the prime denotes the derivative with respect to variable Q. The conditions of the hyperline of the Lifshitz point are just equations (b) in Eq. 59 where Q is put at zero.

Theoretical analysis revealed [22] that the spinodal of a melt of protein-like heteropolymers, as well as that of Markovian multiblock copolymers, has exclusively the trivial branch. Its equation

$$2\chi_{\mathrm{sp}} = H(0) = (l\sigma_l^2)^{-1} \tag{60}$$

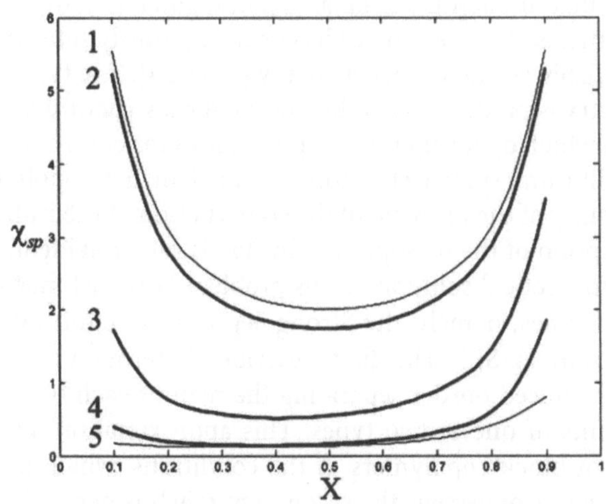

Fig. 4 Curves χ_{sp} versus copolymer composition $X = X_1$ for macromolecules of length $l = 10^3$ at values of parameters (Eq. 20) equal to $b = 0.01$ and $h = 2$ (2); 5 (3); 15 (4). Asymptotic dependences at $h \to 0$ and $h \to \infty$ are depicted by *thin lines* (Eq. 1) and (Eq. 5), respectively. Volume fraction of monomeric units in a globule $\Phi_{\mathrm{g}} = 0.9$

permits finding the spinodal value χ_{sp} of the Flory–Huggins parameter at which the incompressible melt of a heteropolymer looses its thermodynamic stability. Knowledge of χ_{sp} enables one at given dependence $\chi(T)$ to specify temperature T_{sp} separating on the phase diagram regions wherein the phase transition proceeds by the nucleation and by the spinodal decomposition mechanism. Value χ_{sp} (Eq. 60) is easy to calculate making use of formula (Eq. 48) for the dispersion of the composition distribution. Since the value of the latter is controlled along with chemical size l and composition X of a macromolecule also by the mode of the polymeranalogous reaction in a globule, the χ_{sp} value is apparently governed by parameters (Eq. 20) of the reaction system. Considering them as being fixed, it is of interest to clarify how χ_{sp} changes with composition of macromolecules of length l. As follows from Fig. 4, the dependence χ_{sp} versus X is asymmetric with respect to point $X = 0.5$, that qualitatively distinguishes proteinlike heteropolymers from Markovian ones. Among other peculiarities of curves $\chi_{sp}(X)$, of particular interest is the tendency of these curves to become increasingly more flat with the growth of the Thiele modulus h while their minimum value substantially decreases.

3.5.3
Phase Diagram

It is well known that in melts and solutions of block copolymers the thermodynamically stable mesophases can form showing spatially-periodic distributions of densities of monomeric units [32]. It is possible to single out among them three classical mesophases with the translational symmetry corresponding to that of one-dimensional (L), hexagonal (H) and body-centered cubic (BCC) lattices. Their existence has been rigorously substantiated theoretically and reliably verified experimentally [5, 33, 34]. The type of mesophase spatial symmetry is predetermined by the chemical structure of a heteropolymer specimen which opens up fresh opportunities for controlling the morphology of equilibrium spatial structures formed in heteropolymer systems. Thus, the actuality of the problem of the construction of their phase diagrams and the description of the mesophases formed is rather evident.

Under the theoretical solution of this problem, it is customary to consider two limiting regimes, namely, the Strong Segregation Limit (SSL) and Weak Segregation Limit (WSL). The first describes heteropolymer liquids with a strongly pronounced border separating the regions each being enriched by monomeric units of one of two types. This approximation describes pretty well the melt of block copolymers in the conditions, which are far enough from critical ones. Conversely, the second approach is perfectly suited to the description of the thermodynamic properties of such a melt in the vicinity of the critical point when the deviation $\Delta\rho_\alpha(r) = \rho_\alpha(r) - \overline{\rho}_\alpha$ of the densities of different types of monomeric units $\rho_\alpha(r)$ at all points r of a mesophase from their average values $\overline{\rho}_\alpha$ is sufficiently small.

The construction of the phase diagram of a heteropolymer liquid in the framework of the WSL theory is based on the procedure of minimization of the Landau free energy $\mathcal{F}$ presented as a truncated functional series in powers of the order parameter with components $\psi_\alpha(r)$ proportional to $\Delta\rho_\alpha(r)$. The coefficients of this series, known as vertex functions, are governed by the chemical structure of heteropolymer molecules. More precisely, the values of these coefficients are entirely specified by the generating functions of the chemical correlators. Hence, before constructing the phase diagram of the specimen of a heteropolymer liquid, one is supposed to preliminarily find these statistical characteristics of the chemical structure of this specimen. Here a pronounced interplay of the statistical physics and statistical chemistry of polymers is explicitly manifested.

Examination of the thermodynamic behavior of heteropolymer liquids in the framework of the WSL approximation rests on the formalism of the Landau theory of phase transitions that is known to be successfully employed in theoretical physics for the description of various equilibrium systems [35]. Noteworthy, this formalism being applied for the description of polydisperse heteropolymers features some essential peculiarities, which are absent in the Landau theory of low-molecular liquids and monodisperse polymers. So, the inhomogeneity in size, composition and chemical structure of macromolecules constituting a specimen calls for the addition of the so-called nonlocal terms into the expansion of its Landau free energy [36–38]. A general algorithm was formulated for heteropolymers of arbitrary chemical structure nearly ten years ago [5]. Later, an original diagram technique was developed [14], which is rather convenient for practical realization of the above-mentioned algorithm as applied to particular systems.

This diagram technique was employed [39] to find vertex functions entering in the expression for the Landau free energy of the incompressible melt of a proteinlike heteropolymer obtained by a polymeranalogous reaction in a dense homopolymer globule. When tackling this problem, they made use of the expressions Eqs. 42, 43, and 54 derived earlier [22], as well as of the idea formulated in work [40]. The key point of this idea is that expanding the Landau free energy of a polydisperse multiblock copolymer, one cannot confine oneself to fourth order terms with respect to $\psi(r)$, but one has to add some sixth order terms. Using the melt of a binary Markovian copolymer as an example, it was convincingly demonstrated [40] that the allowance for these terms appreciably affects the appearance of the phase diagram. Just the same inference was made under theoretical study of the thermodynamic behavior of proteinlike heteropolymers [39]. Of obvious interest is the comparison of the phase diagrams of these two classes of polydisperse heteropolymers calculated in the framework of the same model.

To carry out such a comparison, the first step is to choose the criterion underlying this procedure. It seems most natural to compare the phase diagrams at the same value of either the average length of blocks (criterion

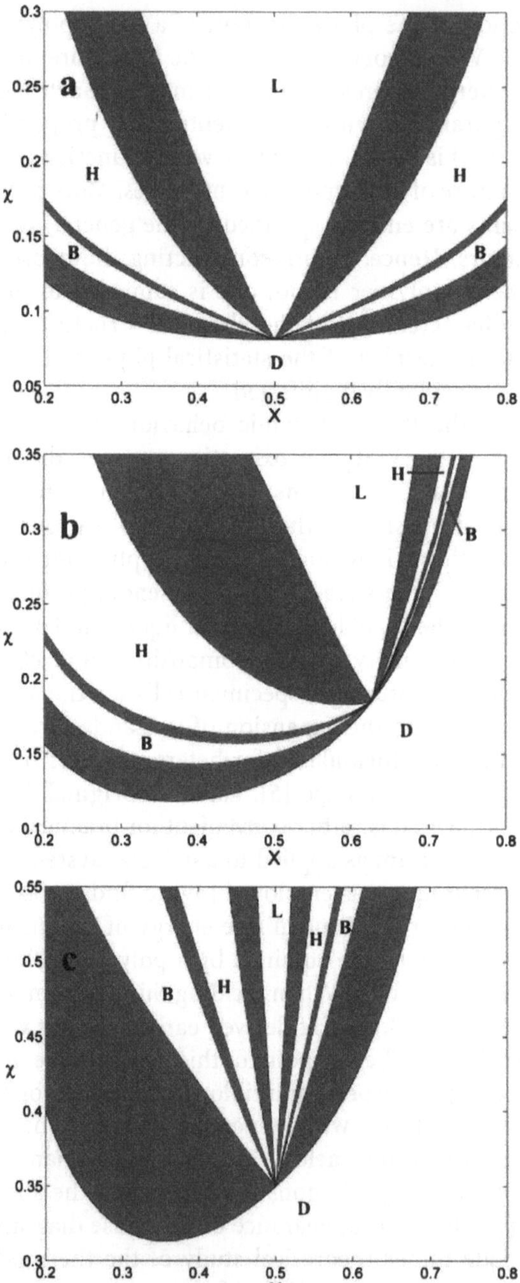

Fig. 5 Comparison of phase diagrams calculated for the melt of a proteinlike heteropolymer (**b**) with the phase diagram of a Markovian copolymer according to criterion II (**a**) and criterion I (**c**). Proteinlike heteropolymer consisting of $l = 10^3$ units is obtained for polymeranalogous reaction in a homopolymer globule at the value of the Thiele modulus h equal to 35

I) or the correlation length n^* (criterion II). In both cases, the sequence of the first order phase transitions under decreasing temperature for protein-like copolymers coincides with the analogous sequence for the Markovian copolymers (see Fig. 5). However, there are substantial quantitative distinctions between them. Firstly, the critical point for Markovian copolymers is observed when the copolymer composition is equimolar. Secondly, at given copolymer compositions, temperatures under which a heteropolymer melt undergoes various transitions, differ in these two cases; and finally, whereas for Markovian copolymers the phase diagram is symmetric with respect to a vertical line passing through the critical point, for melts of proteinlike copolymers such symmetry is absent. The last assertion is true, provided that criterion II is employed. When criterion I is used, the phase diagram for Markovian copolymers is asymmetric as well, that is due to the asymmetry in values of the average length of blocks.

Analyzing theoretically the thermodynamic behavior of the melt of a proteinlike copolymer, authors of work [39] did not confine themselves to the construction of its phase diagram. They also calculated the temperature dependencies of amplitudes and periods of mesophases, as well as their volume fractions in two-phase regions on these diagrams. This permitted them to reveal some important distinctions in the thermodynamic behavior of melts of Markovian and proteinlike heteropolymers.

4 Free-Radical Copolymerization in Dilute Solution

4.1 Kinetic Models

The role of reactive centers in this process is performed by free radicals whose chemical reaction with double bonds in monomer molecules leads to the growth of a polymer chain. Below we will address exclusively reaction systems representing a dilute solution of such chains either in monomers or in a mixture with some solvent. Among the first type of systems is a bulk copolymerization at low conversion of monomers ($p \ll 1$), whereas to the second type a solution copolymerization at arbitrary conversion p pertains, provided the initial molar fraction of monomers is small in comparison with that of a solvent. In systems of both types, polymer chains propagate independently of one another, so that each macroradical may be envisaged as an individual "microreactor" with boundaries permeable to monomer molecules. If the rate of their diffusion V_{diff} is appreciably faster than the rate V_{chem} of the addition to the growing macroradical, the configurational statistics of macromolecules being formed under the condition of the applicability of the ideal kinetic model will be described by the Markov chain [1–4, 41]. Elements of its transition matrix in a simple way

$$\nu_{12} = \frac{x_2}{r_1 x_1 + x_2} = 1 - \nu_{11} \qquad \nu_{21} = \frac{x_1}{x_1 + r_2 x_2} = 1 - \nu_{22} \tag{61}$$

are expressed through molar fractions x_1, x_2 of monomers M_1,M_2 in the microreactor and their reactivity ratios r_1, r_2. These kinetic parameters just slightly depend on temperature and solvent, while their values are tabulated for hundreds of pairs of monomers [42]. Having substituted expression (Eq. 61) for ν_{12} and ν_{21} into formulas Eqs. 8–13, we will find the dependencies of the main statistical characteristics of the chemical structure of polymer chains on local composition of monomer mixture $x = (x_1, x_2)$ in the microreactor.

When examining the initial stages of copolymerization in the framework of the traditional approach, the local composition x is taken to be equal to its global value throughout the whole reaction system. Such an assumption is certain to hold, provided the growing macroradical is in a coil conformational state.

This assumption is implicitly present not only in the traditional theory of the free-radical copolymerization [41, 43, 44], but in its subsequent extensions based on more complicated models than the ideal one. The best known are two types of such models. To the first of them the models belong wherein the reactivity of the active center of a macroradical is controlled not only by the type of its ultimate unit but also by the types of penultimate [45] and even penpenultimate [46] monomeric units. The kinetic models of the second type describe systems in which the formation of complexes occurs between the components of a reaction system that results in the alteration of their reactivity [47–50]. Essentially, all the refinements of the theory of radical copolymerization connected with the models mentioned above are used to reduce exclusively to a more sophisticated account of the kinetics and mechanism of a macroradical propagation, leaving out of consideration accompanying physical factors. The most important among them is the phenomenon of preferential sorption of monomers to the active center of a growing polymer chain. A quantitative theory taking into consideration this physical factor was advanced in paper [51].

4.2
Phenomenon of Preferential Sorption

For the first time attention to the highly important role played by the thermodynamic factors in the formation of macromolecules during copolymerization was drawn almost a quarter of a century ago [52]. When investigating the copolymerization of styrene with methacrylic acid in a solution of CCl_4 and in a solution of dioxane in the region of low conversions, the authors established that copolymers with the same composition had an identical microstructure regardless of the solvent type and of the monomer molar ratio in the feed. This finding led them to the conclusion that the same mechanism and the same set of reactivity ratios were applicable for both systems, despite the substantial difference in the shape of the curves depicting the copolymer composition versus monomer feed composition [52]. Later Harwood [53] extended this conclusion having included into consideration other solvents (benzene and DMF) as well as the systems where instead of acrylic acid another polar monomer (methacrylic acid or methacrylinitrile) participates in the copolymerization with styrene. The presence of the above phenomenon has been reported [54–60] for a number of particular systems of solution copolymerization of vinyl monomers.

It was in article [52] where the main reason responsible for the abovementioned peculiarities was explicitly formulated and substantiated. Its authors related these peculiarities with partitioning of monomer molecules between the bulk of a reaction mixture and the domain of a growing polymer radical. This phenomenon induced by preferential sorption of one of the monomers in such a domain is known as the bootstrap effect. This term was introduced by Harwood [53], because when growing a polymer radical can control under certain conditions its own microenvironment. This original concept enabled him to interpret many interesting features peculiar to this phenomenon. Particularly, he managed to qualitatively explain the similarity of the sequence distribution in copolymerization products of the same composition prepared in different solvents under noticeable discrepancies in composition of monomer mixtures.

The main peculiarities of the preferential sorption phenomenon have been experimentally examined in detail in a series of papers by Semchikov, Smirnova et al. [61–76]. Investigating the free-radical copolymerization in bulk of about 30 concrete pairs of the most commonly encountered vinyl monomers, they revealed the following characteristic features of this phenomenon:

1) The composition of a number of copolymers synthesized in the region of low conversions is markedly controlled by their molecular weight within a particular area of its values.
   Essentially, at a fixed monomer feed composition and temperature such a dependence is of universal character irrespective of the way of molecular weight regulation, which could be performed by changing the concentration either of the initiator or the transfer agent. Just on this universal curve the points fall which characterize the composition and molecular weight of the fractions isolated from copolymers synthesized under different contents of initiator and transfer agent.
2) For the copolymerization products obtained under the conversion less than 10 percent exhibit the composition inhomogeneity substantially exceeding that described by the traditional theory of free-radical copolymerization.

The above phenomenon is due to the pronounced polydispersity of these products in their chemical size l described by the Flory exponential distribution. Because the composition of each macromolecule of the sample under investigation is unambiguously related to its degree of polymerization l, the Flory distribution for l in a polymer sample is responsible for its significant composition inhomogeneity.

3) Each copolymer macromolecule is chemically inhomogeneous along its length.

This means, that macromolecular fragments of the same length positioned in different parts of the polymer chain differ in composition. According to the concepts of the traditional theory of free-radical copolymerization, such intramolecular inhomogeneity is negligible, since a polymer radical grows under the condition of a constant monomer feed composition which actually has no chance to change during its lifetime. This statement does not obviously hold if the phenomenon of preferential sorption manifests itself, i.e., when the local monomer composition in the vicinity of the growing radical drifts along with the macroradical composition at constant global monomer composition in bulk. In order to reveal the intramolecular inhomogeneity experimentally, it would be natural to proceed from the fact that the macromolecules exhibiting this property decompose to yield products, which due to the difference in composition favor the broadening of the composition distribution of a copolymer sample. In the absence of the intramolecular inhomogeneity, such a distinction will not be the case, and thus the analysis of the broadening of composition distribution of the products being formed at the initial stage of the destruction provides a possibility to draw a conclusion about the degree of such an inhomogeneity. Its presence has been conclusively established [74–76] at the qualitative level just in such a way.

The systems where the above-discussed peculiarities were observed have been termed anomalous [70, 75] to differentiate them from classical ones, describable by the traditional theory of free-radical copolymerization. The proportion of the latter among 32 systems examined in the series of works [61–76] constitutes about one third. Among the remaining two thirds, the dependence has been noted of the copolymer composition on the concentration of the initiator or chain transfer agent, while in more than half of such anomalous systems this dependence was found to be rather strongly pronounced. Hence, the phenomenon of the preferential sorption of monomers by growing macroradicals is of fairly general character. Consequently, when developing a theory of free-radical copolymerization enabling a quantitative description of its peculiarities formulated above, the necessity of the allowance for this phenomenon is beyond question. The results achieved in paper [51] where for the first time an attempt was undertaken to elaborate such a theory based on the Harwood Bootstrap Model [53] and current concepts of statistical physics and statistical chemistry of polymers is discussed below.

4.3
Background of the Physicochemical Model

At the initial stage of a bulk copolymerization, a reaction system represents a dilute solution of heteropolymer in a mixture of its monomers when macromolecules are widely spaced. Under such conditions, each macroradical is a separate microreactor with the boundaries transparent for monomer molecules. Their concentrations in this microreactor are governed by the thermodynamic equilibrium, because the reaction of a polymer chain propagation for the system in point is kinetically controlled. This statement is easy to verify having compared the characteristic time of the diffusion of a monomer molecule inside the microreactor and that of its addition to the macroradical. The latter is, as a rule, several orders of magnitude higher than the former, so that the entry of monomer molecules into the chemical reaction does not virtually affect their equilibrium partitioning between the bulk of the reaction system and the domains of growing polymer radicals. Finding of this distribution constitutes one of the two main problems whose solution is indispensable for the elaboration of a copolymerization theory that allows for the preferential sorption of monomers. Let us discuss the assumptions, which have been made [51] to cope with this problem.

With the growth of a number of units l of a polymer chain, their density in a microreactor is expected to approach either zero or to remain non-vanishing, depending on whether this chain is in a coil or globule state. In the first case this density at $l \sim 10^3$–10^4 is so small, that a pronounced distinction between the concentration of monomers in the microreactor and their values in bulk can hardly be expected. This has been convincingly supported by simple estimates by Smirnova [76], based on her experimental data, obtained for the examination of copolymerization in bulk of an equimolar mixture of styrene and methacrylic acid. By virtue of two independent methods (equilibrium dialysis and the light scattering), it has been established that the enhancement of the copolymer molecular weight from 3×10^4 to 1×10^6 is accompanied by the growth of the coefficient of preferential sorption of methacrylic acid by $\Delta\lambda = 0.35$ ml/g under simultaneous increase of the content of its units in macromolecules by 10–12 molar percent. To such an alteration in copolymer composition ΔX there corresponds the change of (15–17%) in the monomer feed composition Δx inside the microreactor. This value exceeds approximately by a factor of ten the value $\Delta x \sim (1$–$2\%)$, which ensues from simple estimates using the value $\Delta\lambda = 0.35$ ml/g on the assumption that a macroradical represents a Gaussian coil. On encountering this contradiction, Smirnova [76] conjectured that a monomer preferentially sorbed by the macroradical is mainly localized in the vicinity of a polymer chain in a solvate layer whose composition differs from that of the monomer mixture in bulk. On the contrary, another concept has been put forward [51] that permits elimination of the above contradiction without recourse to the assumption about the decisive role of adsorption of monomers on polymer chains.

In this paper it was stated that the effect of preferential sorption of monomers, sufficient for noticeable change of their composition in a microreactor, shows up only when the growing macroradical is in a state of a globule. Inside it, the density of monomeric units depending on their thermodynamic affinity to the solvent may have a wide range of values, in particular, those substantially exceeding the value, which could have a macroradical of large chemical size in a coil state. A considerable body of experimental data regarding the preferential sorption of binary solvent molecules by different homopolymers points to a dramatic intensification of this effect in the proximity of the Θ-point [77, 78]. At this particular point, by its definition [79], a macromolecule with $l \to \infty$ undergoes a transition from a coil state to a globular one. Under further deterioration of thermodynamic affinity between the polymer and the solvent, such a transition happens for polymer chains of finite chemical size l, which is the smaller, the lower this affinity. As one moves away from the Θ-point, the increase in polymer density inside the core of a globule is accompanied by narrowing of the thickness of the surface layer on its boundary. The difference of free energy of a globule and a coil comprises volume and surface contributions, having for thermodynamically poor solvent different signs. The coil-to-globule transition takes place when these contributions become equal in absolute value. Traditionally [79], the cases are examined when such a transition in a macromolecule of fixed length is induced by the change of either the temperature or the thermodynamic quality of the solvent. On the contrary, in this review we are concerned with another case where the above external conditions remain unaltered, but the chemical size l of a macromolecule is enhanced.

The following explanation was suggested [51] of the peculiarities due to the preferential sorption phenomenon which are normally observed during the initial stage of free-radical copolymerization. According to this explanation, such peculiarities noticeably manifest themselves when a monomer mixture is a moderately poor solvent for the copolymer chains formed. If the solvent medium exhibits lower affinity to the polymer, it will behave as a precipitating agent, thus conditioning the heterophase regime of copolymerization. Conversely, under higher affinity when the monomer mixture is a good solvent, strongly pronounced anomalies inherent in the free-radical copolymerization are not likely to arise.

All three regimes discussed in the foregoing have been observed experimentally [62] under bulk copolymerization of styrene and methacrylic acid for some specific ranges of monomer mixture composition whose thermodynamic affinity toward the copolymer was found to progressively decrease while increasing the acid fraction. In this system, the dependence of the composition of a copolymer on its molecular weight is most strongly pronounced in the boundary area around the region of the heterophase regime of copolymerization. This behavior, featured also by other anomalous systems [62, 76], agrees with the hypothesis concerning the role of a globular state of macroradicals advanced by authors of paper [51]. The explanation is that in dilute solutions the coil-to-globule transition because of the worsening of the affinity between polymer and solvent is immediately followed by polymer precipitation as a separate macrophase [79]. The distinction in temperatures under which these two thermodynamic transitions occur is due to the translational entropy of macromolecules.

By adding to the monomer mixture an extra low-molecular component, one can materially either increase or decrease its solvent power. This method of controlled alteration of thermodynamic affinity between solvent and polymer provides a possibility to verify our globular state hypothesis using the experimental data earlier obtained by Smirnova, Semchikov et al. [62, 76]. They have chosen one typical representative of the classical (styrene + methylmethacrylate) and one of the anomalous (styrene + methacrylic acid) systems and carried out copolymerization of the first of these monomer pairs in the presence of the precipitating agent (cyclohexane or methyl alcohol), while the copolymerization of the second pair was conducted in the presence of dimethyl formamide, which is a good solvent. It was established that in the first system the addition of the precipitating agent to the extent insufficient for phase separation results in the appearance of a considerable dependence of composition of a copolymer on its molecular weight. On the other hand, in the second system, the addition of the solvent was found to lead to the disappearance of this effect. Consequently, its appearance signals that all copolymer macromolecules have been formed in the medium of the moderately poor solvent.

Under such conditions, a polymer chain begins growing in a coil state, and then, once a certain critical value of l^* units is attained, it changes into a globular state, continuing its growth up to the moment when the chain is terminated by interaction with either another radical or with a chain transfer agent. The most interesting case will now be discussed, where l^* is far less than the average chemical size, i.e., for most of its lifetime the macroradical grows in a globular state. Just in this case the most pronounced anomalies should be expected as compared to the classical picture of radical copolymerization. The reason for such anomalies resides in the fact that the monomer mixture composition in a globule, because of the phenomenon of preferential sorption, differs, generally speaking, from its value in bulk. Moreover, thermodynamic equilibrium values of the monomer concentrations in a globule are not likely to be constant during the process of macroradical growth, but will change as a consequence of the alteration of the propagating copolymer chain composition. Since the latter by itself depends on the monomer mixture composition in the microreactor, an interplay between physical and chemical factors is the case here by virtue of which the propagating copolymer radical can control its own environment.

Such an interplay is reflected in the right-hand part of equation

$$\frac{dX_\alpha}{d\xi} = \pi_\alpha(x) - X_\alpha \quad \text{where} \quad \xi = \ln l \qquad (\alpha = 1, 2) \tag{62}$$

describing the evolution of the chemical composition of a growing macroradical with the increase of its length l. The dependence of components $\pi_\alpha(\boldsymbol{x})$ of the vector $\boldsymbol{\pi}(\boldsymbol{x})$ on the monomer mixture composition $\boldsymbol{x}$ in the case of the ideal kinetic model is defined by formulas (Eqs. 8 and 61). Equation 62, resulting from the material balance conditions should be complemented by equations

$$x_\alpha = F_\alpha(X) \qquad (\alpha = 1, 2) \tag{63}$$

describing the dependence of the vector $\boldsymbol{x}$ on $\boldsymbol{X}$ in the state of thermodynamic equilibrium. To find vector-function $\boldsymbol{F}(\boldsymbol{X})$ it is obviously necessary to solve the thermodynamic problem on equilibrium partitioning of monomers between globules and the solution surrounding them.

4.4 Thermodynamics of Preferential Sorption

Under the theoretical description of this phenomenon, the authors of paper [51] considered globules of the traditional morphology. Most of such a globule is occupied by a spatially homogeneous core where the volume fraction of each of the low-molecular weight components φ_α (monomers) and quasi-components Φ_α (monomeric units) has a definite value identical all over the core. Only within a rather narrow surface layer of the globule do the volume fractions change from φ_α and Φ_α in the globule core to φ_α^0 and 0 in the surrounding solution. Examination of this layer, necessary to find the surface tension at the globule boundary, constitutes a far more complicated task in comparison with the description of the globule core. The reason for these difficulties, for the calculation of concentration distributions of the components and quasi-components within the surface layer, has to do with the fact that these distributions are prescribed by the type of arrangement of monomeric units along a macromolecule, whereas the values of these concentrations in the globule core are not affected by such an arrangement.

Volume approximation (when the surface contribution to the free energy of a globule is neglected) works the better the farther the system is from the point of the coil-to-globule transition. In the framework of this approximation, it coincides with the Θ-point, whereas under the theoretical consideration where the surface layer is taken into account, a gap appears separating these two points. The less is the length of polymer chain l, the more pronounced is this gap. Hence, the condition, imposed on the thermodynamic and stoichiometric parameters of the system by the equation of the Θ-point, specifies the position of the boundary separating that part of the space of these parameters where the bootstrap effect should be expected.

From a thermodynamic point of view, the heteropolymer globule in hand represents a subsystem which is composed of a macromolecule involving l_1, l_2 units $\overline{M}_1, \overline{M}_2$ and molecules of monomers M_1, M_2 whose numbers are N_1, N_2. Among these variables and volume fractions $\varphi_\alpha, \Phi_\alpha$ in the framework of the simplest Flory–Huggins lattice model there are obvious stoichiometric relationships

$$X_\alpha = \frac{l_\alpha}{l} \qquad x_\alpha = \frac{N_\alpha}{N_1 + N_2} \qquad \Phi_\alpha = \frac{l_\alpha}{N} = \Phi X_\alpha \qquad \varphi_\alpha = \frac{N_\alpha}{N} = \varphi x_\alpha \tag{64}$$

$$\Phi = \Phi_1 + \Phi_2 = \frac{l}{N} \qquad \varphi = \varphi_1 + \varphi_2 = \frac{N_1 + N_2}{N} \qquad N = l + N_1 + N_2$$

using which it is an easy matter to write down the expression for the free energy of the globule under consideration

$$\frac{G}{NT} = \sum_{\alpha=1}^{2} \varphi_\alpha \ln \varphi_\alpha + h(\varphi, \Phi) . \tag{65}$$

The first term in the right-hand part of the expression (Eq. 65) corresponds to the translational entropy of low-molecular weight components, whereas the second one involves three contributions

$$h(\varphi, \Phi) = h^{\mathrm{MM}} + h^{\mathrm{MP}} + h^{\mathrm{PP}} , \quad h^{\mathrm{MP}} = h^{\mathrm{PM}} = \Phi \sum_{\alpha=1}^{2} \sum_{\beta=1}^{2} \chi_{\alpha\beta}^{\mathrm{PM}} X_\alpha \varphi_\beta \tag{66}$$

$$h^{\mathrm{MM}} = \frac{1}{2} \sum_{\alpha=1}^{2} \sum_{\beta=1}^{2} \chi_{\alpha\beta}^{\mathrm{MM}} \varphi_\alpha \varphi_\beta , \quad h^{\mathrm{PP}} = \frac{1}{2} \sum_{\alpha=1}^{2} \sum_{\beta=1}^{2} \chi_{\alpha\beta}^{\mathrm{PP}} X_\alpha X_\beta ,$$

which describe the enthalpy of mixing (per site of lattice) of these components: (i) one with the other (h^{MM}), (ii) with monomeric units of a macromolecule (h^{MP}), as well as that of units between themselves (h^{PP}). Along with the Flory–Huggins parameters $\chi_{\alpha\beta}$, the expression (Eq. 65) also includes the absolute temperature T (expressed in energetic units), the number of sites N contained in a polymer globule volume and the fraction $\Phi = l/N$ of this volume occupied by monomeric units.

Equilibrium values of numbers N_α of monomer molecules M_α in a globule can obviously be found from the condition of the equilibrium of the values of the chemical potentials of these molecules inside, μ_α, and outside, μ_α^0, this globule. The explicit expressions for μ_1 and μ_2 are obtained by differentiation of the relationship (Eq. 65), complemented by the incompressibility condition $\varphi + \Phi = 1$, with respect to N_1 and N_2, respectively. With these expressions at hand, the set of two equations, ($\mu_1 = \mu_1^0, \mu_2 = \mu_2^0$), for the calculation of N_1 and N_2 has been derived [51]. Having these quantities calculated, it is easy to find the unknown dependence (Eq. 62) of the monomer mixture composition $\boldsymbol{x}$ on copolymer composition $\boldsymbol{X}$. Such a procedure provides the possibility of establishing an equilibrium correlation between vectors $\boldsymbol{x}$ and $\boldsymbol{X}$ which, in conjunction with kinetic equations (Eq. 62), enables one, in terms of the model chosen, to describe macroradical growth with allowance for preferential sorption of monomers in its domain.

4.5 Quantitative Theory of Copolymerization in Anomalous Systems

As pointed out in the foregoing, there are two specific peculiarities qualitatively distinguishing these systems from the classical ones. These peculiarities are intramolecular chemical inhomogeneity of polymer chains and the dependence of the composition of macromolecules $\boldsymbol{X}$ on their length l. Experimental data for several nonclassical systems indicate that at a fixed monomer mixture composition $\boldsymbol{x}^0$ and temperature such dependence of $\boldsymbol{X}$ on l is of universal character for any concentration of initiator and chain transfer agent [63, 72, 76]. This function $\boldsymbol{X}(l)$, within the context of the theory proposed here, is obtainable from the solution of kinetic equations (Eq. 62), supplemented by thermodynamic equations (Eq. 63). For heavily swollen globules, when $\Phi_G \ll 1$, the components of the vector-function $\boldsymbol{F}(\boldsymbol{X})$ can be presented in explicit analytical form

$$x_\alpha = F_\alpha(\boldsymbol{X}) = x_\alpha^0 + \kappa_\alpha \Phi_G \qquad (\alpha = 1, 2)\,, \tag{67}$$

where the dependence on $\boldsymbol{X}$ of the values Φ_G and κ_α, characterizing the total and the preferential sorption of monomers in the microreactor, respectively, is obtainable from the thermodynamic considerations [51].

In nonclassical systems, a polymer chain growing in a coil state, once attaining the length $l = l^*$, turns into a globule in which the monomer mixture composition $\boldsymbol{x}' = \boldsymbol{F}(\boldsymbol{x}^0)$ (where $\boldsymbol{X}^0 = \boldsymbol{\pi}(\boldsymbol{x}^0)$) differs, generally speaking, from that, $\boldsymbol{x}^0$, observed in bulk. This leads to the evolution of the growing macroradical composition described by Eq. 62 and thus to the drift of the monomer mixture composition in the microreactor in accordance with the dependence (Eq. 63). Here the obvious question arises to which limiting values $\boldsymbol{X}^\infty$ and $\boldsymbol{x}^\infty$ the above compositions will tend at $l \to \infty$. A general answer to this question is more readily attained by considering this process in terms of the theory of dynamic systems, just along the same lines as has been done earlier [1, 3, 80] when dealing with classical multicomponent copolymerization within the region of high conversions.

For the case of interest, copolymerization dynamics is described by nonlinear equations (Eq. 62) where variable ξ plays the role of time, supplemented by the thermodynamic relationship (Eq. 63). The instantaneous state of the system characterized by vector $\boldsymbol{X}$ may be represented by a point inside the unit interval $X_1 + X_2 = 1$. The evolution of composition $\boldsymbol{X}$ in the course of a polymer chain growth can be described by the movement of a point, characterizing the system state, inside this interval. Its position within the unit interval is governed by the value of the variable ζ. The knowledge of such a trajectory is of prime importance, since it controls the main statistical characteristics of the copolymer formed. Thus, its average composition can be calculated by formula

$$\langle X_\alpha \rangle = \int_1^\infty f_W(l) X_\alpha(l) \mathrm{d}l , \tag{68}$$

where angular brackets mean that the value inside is averaged over the weight distribution of macromolecules $f_W(l)$ for their chemical size l. In order to find the weight composition distribution $f_W^c(\zeta)$, the Dirac delta-function should be analogously averaged

$$f_W^c(\zeta) = \langle \delta\left[\zeta - \boldsymbol{X}(l)\right] \rangle \tag{69}$$

because in this case the composition inhomogeneity of macromolecules of the same chemical size l is neglected.

At low-conversion copolymerization in classical systems, the composition of macromolecules $\boldsymbol{X}$ whose value enters in expression (Eq. 69) does not depend on their length l, and thus the weight composition distribution $f_W^c(\zeta)$ (Eq. 1) equals $\delta(\zeta - \boldsymbol{X}^0)$ where $\boldsymbol{X}^0 = \boldsymbol{\pi}(\boldsymbol{x}^0)$. Hence, according to the theory, copolymers prepared in classical systems will be in asymptotic limit $\langle l \rangle \to \infty$ monodisperse in composition. In the next approximation in small parameter $1/\langle l \rangle$, where $\langle l \rangle$ denotes the average chemical size of macromolecules, the weight composition distribution $f_W^c(\zeta)$ will have a finite width. However, its dispersion specified by formula (Eq. 13) upon the replacement in it of l by $\langle l \rangle$ will be substantially less than the dispersion of distribution (Eq. 69)

$$\sigma^2 = \langle X_1^2 \rangle - \langle X_1 \rangle^2 = \langle X_2^2 \rangle - \langle X_2 \rangle^2 . \tag{70}$$

The solution to the problem of sequence distribution as well as that of composition distribution (Eq. 69) reduces to the calculation of simple integrals. So, the probability $P\{U\}$ of an arbitrary sequence $U = \{\overline{M}_\alpha \overline{M}_\beta \cdots \overline{M}_\psi \overline{M}_\omega\}$ in macromolecules of a copolymer can be calculated by the following formula

$$P\{U\} = \left\langle \frac{1}{l} \int_1^l \pi_\alpha[\boldsymbol{x}(l')] \nu_{\alpha\beta}[\boldsymbol{x}(l')] \cdots \nu_{\psi\omega}[\boldsymbol{x}(l')] \mathrm{d}l' \right\rangle , \tag{71}$$

where the dependence of the components π_α of the stationary vector $\boldsymbol{\pi}$ and elements $\nu_{\alpha\beta}$ of the matrix of transition probabilities on monomer mixture composition is given by formulas (Eqs. 8 and 61).

For the calculation of the statistical characteristics of a copolymer by means of expressions (Eqs. 68–71), it is necessary to substitute into them the distribution function $f_W(l)$ for the chemical size of macromolecules formed under initial conversions. As such a function one may choose the Flory distribution [3]

$$f_W(l) = \left[\frac{1-\gamma}{2}\frac{l}{l_{av}} + \gamma\right]\frac{l}{l_{av}^2}\exp\left\{-\frac{l}{l_{av}}\right\} \tag{72}$$

involving two parameters, which are the fraction γ of radicals terminating by disproportionation and the average length of kinetic chain l_{av}. It is expressed in a simple manner $l_{av} = (1+\gamma)P_N/2 = P_W/(3-\gamma)$ through number-average P_N or weight-average P_W degree of polymerization. Expression (Eq. 72) has been derived for a polymerization in the absence of a chain transfer agent, which, if present, does not affect the form of the function $f_W(l)$ being responsible only for the re-definition of its parameters [3]. The latter, along with reactivity ratios r_1, r_2, monomer feed composition $\boldsymbol{x}^0$ and the Flory–Huggins parameters $\{\chi_{\alpha\beta}\}$, constitute a complete set of control parameters which in terms of the proposed theory are indispensable for calculating the statistical characteristics of the chemical structure of copolymers in nonclassical systems.

5 Interphase Free-Radical Copolymerization at the Oil–Water Boundary

5.1 Mechanism of Copolymerization

It is well known that under a free-radical copolymerization conducted in the traditional way, macromolecules are formed whose units are distributed statistically. This result does not depend on whether polymer chains are formed in bulk, solution or inside latex particles. The impossibility to obtain multiblock copolymers in the course of a copolymerization carried out in a homophase medium, stems immediately from the examination of the kinetics of the propagation of a polymer chain. In fact, for the long blocks of single type units to be formed, a macroradical with terminal unit $\overline{M}_\alpha$ should add a monomer of the same type, M_α, with much higher probability than a monomer of the other type, M_β. In other words, the formation of a multiblock copolymer in a homophase system occurs only provided both reactivity ratios (Eq. 61) have sufficiently large values $r_1 \gg 1$ and $r_2 \gg 1$. However, among many experimentally studied copolymerization systems [88], none has been found meeting these conditions. This brings up the question as to whether multiblock copolymers can be synthesized at all by the free-radical mechanism. The answer turns out to be positive [71]. Indeed, such a possibility exists, provided one of the monomers involved in the copolymerization is hydrophilic, while the other one is hydrophobic.

In this case, a reaction system represents an oil–water miniemulsion with two monomers solved, respectively, inside and outside the droplets of this miniemulsion. In such a system the growth of a polymer chain may occur in the vicinity of the boundary separating two immiscible liquid phases, each containing molecules of only one type of monomers. In this case, an active center of a growing polymer radical can cross the interface in both directions, which leads to the formation of polymer chains with each type of monomeric unit arranged in blocks (Fig. 6). Such block copolymer macromolecules showing an extraordinary high surface activity will be disposed exclusively near the interface acting, in essence, as a polymer emulgator.

The aforementioned mechanism of the growth of polymer chains will be realized, provided a surface-active initiator (SAI) is involved. Conversely, if copolymerization is conducted in the presence of a traditional initiator, solvable in either of two phases, the only products of such a process will be homopolymer molecules. It should be attributed to the fact that a homopolymer chain is not a surfactant, and, consequently, the probability for the terminal monomeric unit of a growing macroradical to fall on the interface is negligible. In the presence of SAI a polymer chain starting its growth close to this boundary will stay later in its vicinity. This is because a SAI fragment adjoined to the inactive end of a homopolymer radical residing on the interfacial surface acts as an "anchor" which prevents this radical from going deep into the volume of one of the phases. If the surface activity of an initiator is sufficiently pronounced, a macroradical remains in the neighborhood of the interface up to the instant when its reactive end crosses this boundary. Thereafter, the growth of the second block of a macroradical begins resulting in the transformation of the latter into a polymer emulgator virtually incapable of abandoning the interfacial area.

Advantageous implementation of the above-described mechanism of the formation of block copolymer molecules is largely predetermined by an adequate choice of SAI. The role of the latter may be performed by ordinary oligomer surfactants upon introducing in their molecules peroxide or

Fig. 6 Schematic representation of a polymer chain growing in the vicinity of the interface

diaso-groups, involved in the traditional initiators of the radical polymerization [72]. Synthesizing diblock copolymers that contain on the ends of their macromolecules the above-mentioned labile functional groups, it is possible to obtain polymer SAI showing appreciably greater surface activity compared to that of SAI prepared from traditional surfactants. The main requirement for the initiator chosen is its minor solubility in both phases. This condition should be necessarily met to exclude the formation of homopolymer molecules far from the interphase boundary.

5.2 Physicochemical Model

Considering theoretically a copolymerization on the surface of a miniemulsion droplet, one should necessarily be aware of the fact that this process proceeds in the heterophase reaction system characterized by several spatial and time scales. Among the first ones are sizes of an individual block and macromolecules of the multiblock copolymer, the radius of a droplet of the miniemulsion and the reactor size. Taking into account the pronounced distinction in these scales, it is convenient examining the macrokinetics of interphase copolymerization to resort to the system approach, generally employed for the mathematical modeling of chemical reactions in heterophase systems [73].

According to this approach, it is necessary to consider first the growth of a polymer chain near the interphase boundary of an individual particle and then to perform the averaging over volumes containing a sufficiently large number of the droplets. Such an averaging leads to the equations reminiscent of those used in the homophase kinetics, but with the parameters allowing for the peculiarities of the macrokinetics in heterophase systems. This quasi-homogeneous approach is a reasonably good approximation for the description of the interphase copolymerization at spatial scales far more large than the radius of the droplets. This is because the rate of the monomers' supply toward the interface appreciably exceeds the rate of their addition to the growing macroradicals. In this case, monomer concentration M'_α may be thought of as identical all over α-th phase ($\alpha = 1, 2$) in which monomer M_α is solved.

This monomer concentration M_α in the formalism of the quasi-homogeneous approximation, unlike M'_α, refers to the whole volume of the two-phase system. The aforementioned quantities are connected by the simple relationship $M_\alpha = y^\alpha M'_\alpha$ where y^α stands for the volume fraction of the α-th phase in miniemulsion. An analogous relation, $R_\alpha = sd_\alpha R'_\alpha$, exists between the concentrations R_α of the α-th type active centers in the entire system and those R'_α in the surface layer of the α-th phase. This layer thickness d_α has the scale of average spatial size of the α-th type block, which hereafter is presumed to be small as compared to the average radius of miniemulsion drops. Apparently, in this case, the curvature of the interphase surface can be neglected and thus the propagation of a macroradical can be considered near the plane. In this approximation, the volume fraction of a system, occupied by the surface layer, is equal to the product of its thickness d_α and factor s that is the interphase surface area per unit volume of a miniemulsion.

To elaborate a theory of interphase copolymerization at an oil–water boundary the necessity arises to consider initially the growth of an individual polymer chain near the surface separating the organic and water phases. By the model introduced in paper [74], molecules of only one of the monomers are presumed to be solved inside either of these two phases. A theoretical examination of the formation of macromolecules turns out here to be substantially simpler, since their chemical structure under such an approximation is the same as that of a traditional block copolymer.

Other assumptions of the physicochemical model [74] are the following:

1) The reactivity of a macroradical is controlled only by the type of its terminal monomeric unit.
2) The rate of establishing of the conformational equilibrium of the propagating polymer chain essentially exceeds the rate of the addition by it of a monomer.
3) Probabilities of equilibrium conformations of polymer chains are describable by the Gaussian statistics.

The first of these assumptions, generally accepted in macromolecular chemistry [1, 3], is correct enough when considering the propagation reaction under copolymerization of the majority of monomers. Simple estimates reported in paper [74] support the correctness of the second assumption. As for the third one, it is true, strictly speaking, only under θ-conditions. The conformational statistics of macromolecules in a thermodynamically good solvent is known [30] to differ from the Gaussian one. Nevertheless, this distinction may hardly influence the qualitative conclusions of the simplest theory of interphase copolymerization. To which extent the account of the excluded volume of macromolecules will affect quantitative results of this theory, may be revealed exclusively by computer simulations.

5.3 Key Points of the Theoretical Approach

The development of a quantitative theory of a free-radical copolymerization implies the derivation of equations for the rate of the monomers' depletion and the statistical characteristics of the chemical structure of macromolecules present in the reaction system at the given conversion p of monomers. Elaborating such a theory one should take into account a highly important peculiarity inherent to any free-radical copolymerization. This peculiarity is that the characteristic time of a macroradical life is appreciably less than the time of the process duration. Consequently, its products represent definitely a mixture of macromolecules formed at different moments, i.e. at different monomer mixture compositions. That is why problems of two kinds are generally encountered when developing a quantitative theory of a free-radical copolymerization. The first are those related to finding *instantaneous* values of the statistical characteristics of the chemical structure of macromolecules formed at a given value of conversion p'. However, of particular practical interest are *average* values of these characteristics describing copolymerization products, which are present in the reactor at conversion p. In order to calculate such characteristics it is necessary to average their instantaneous values over all conversions $p' < p$, preliminarily having found the dependence of monomer mixture composition on p'. The realization of such an averaging procedure belongs to the problems of the second kind.

The solution of the problems of the first kind implies an in depth consideration of the formation of a multiblock copolymer macromolecule at the interphase boundary. This process may be described in the framework of the above-discussed physicochemical model as follows.

Upon falling into the α-th phase, the active center situated on the end of this macroradical begins to add monomer M_α until going to the other phase or terminating. The rate of such an addition $\theta_\alpha = k_{\alpha\alpha}M'_\alpha$ is equal to the product of the rate constant $k_{\alpha\alpha}$ of the reaction of a homopolymer chain propagation and concentration M'_α of monomer M_α in the α-th phase. In a time t_α, elapsed between two successive crossings of the α-th phase boundary by the active end of a macroradical, the latter increases its chemical size by the length $l_\alpha = \theta_\alpha t_\alpha$ of a single block of α-th type units. Due to the proportionality of quantities l_α and t_α the chemical structure of a macroradical may be exhaustively specified not only by the sequence of the constituent blocks but also by an analogous sequence of the residence times of an active center in different phases.

Thus, the problem on the growth of a block copolymer chain in the course of the interphase radical copolymerization may be formulated in terms of a stochastic process with two regular states corresponding to two types of terminal units (i.e. active centers) of a macroradical. The fact of independent formation of its blocks means in terms of a stochastic process the independence of times t_α of the uninterrupted residence in every α-th stay of any realization of this process. Stochastic processes possessing such a property have been scrutinized in the Renewal Theory [75]. On the basis of the main ideas of this theory, the set of kinetic equations describing the interphase copolymerization have been derived [74].

So, the concentrations of the radicals of length l with type 1 or 2 terminal units are determined from the following expressions

$$R_1(l) = \int_0^l Q_1(l-\xi)w_1(\xi)\mathrm{d}\xi\theta_1^{-1} \qquad R_2(l) = \int_0^l Q_2(l-\xi)w_2(\xi)\mathrm{d}\xi\theta_2^{-1}\,. \tag{73}$$

Here, function $Q_\alpha(\xi)$, $(\alpha = 1, 2)$ having a meaning of the rate of generating of macroradical with length ξ and α-th type terminal unit, is obtained from the solution of two coupled linear equations

$$Q_1(\xi) = \int_0^\xi Q_2(\xi - \eta) w_2(\eta) V_{21}(\eta) d\eta \theta_2^{-1} + I_1 \delta(\xi) \tag{74}$$

$$Q_2(\xi) = \int_0^\xi Q_1(\xi - \eta) w_1(\eta) V_{12}(\eta) d\eta \theta_1^{-1} + I_2 \delta(\xi) .$$

The kernels of these integral equations, which are derived from simple probabilistic considerations, represent up to the factor θ_α^{-1} the product of two factors. The first of them, $w_\alpha(\eta)$, is equal to the fraction of α-th type blocks, whose lengths exceed η. The second one, $V_{\alpha\beta}(\eta)$, is the rate with which an active center located on the end of a growing block of monomeric units M_α with length η switches from α-th type to β-th type under the transition of this center from phase α into phase β. The right-hand side of Eq. 74 comprises items equal to the product of the rate of initiation I_α of α-th type polymer chains and the Dirac delta function $\delta(\xi)$.

The propagation of a α-th type block of a macroradical may be interrupted either because of the addition of monomer M_β or owing to the loss of an active center caused by the chain termination reaction. The probabilities of these events within the interval $d\tau_\alpha = dl/\theta_\alpha$ are equal to $V_{\alpha\beta}(l)d\tau_\alpha$ and $T_\alpha d\tau_\alpha = k_{t\alpha} R'_\alpha d\tau_\alpha$, respectively. Hereafter, $k_{t\alpha}$ is the constant of the chain termination reaction while R'_α stands for the concentration of α-th type active centers in the surface layer of the α-th phase. Function $w_\alpha(\eta)$, having the sense of the probability for a α-th type terminal block of a macroradical to attain length η, reads as

$$w_\alpha(\eta) = \exp\left\{ - \int_0^\eta \frac{V_{\alpha\beta}(\xi)}{\theta_\alpha} d\xi - \frac{T_\alpha}{\theta_\alpha}\eta \right\} , \tag{75}$$

where indices $\alpha \neq \beta$ run values 1 and 2.

Having hypothetically assumed that rates $V_{12}(\xi)$ and $V_{21}(\xi)$ of an active center transition through the interface do not depend on length ξ of the growing terminal block of a macroradical, one will find the distribution of blocks for length (Eq. 75) to be exponential. In this unreal case, the solution of Eqs. 73 and 74 will formally reduce to the solutions of the traditional equations of radical copolymerization [76] for the concentrations $R_\alpha(l)$ of radicals with length l

$$\theta_1 \frac{dR_1(l)}{dl} = - V_{12}R_1(l) + V_{21}R_2(l) - T_1R_1(l) + I_1\delta(l) \qquad (76)$$

$$\theta_2 \frac{dR_2(l)}{dl} = - V_{21}R_2(l) + V_{12}R_1(l) - T_2R_2(l) + I_2\delta(l)$$

in which $V_{\alpha\beta} = k_{\alpha\beta}M'_\beta$ has the meaning of the rate of the addition of monomer M_β to the α-th type active center.

For the process of the interphase copolymerization, however, the dependence of coefficients $V_{12}(\xi)$ and $V_{21}(\xi)$ of Eq. 74 on ξ proves to be rather substantial. It is determined by the following expressions

$$V_{12}(\xi) = p_1(\xi)k_{12}M'_2 \qquad V_{21}(\xi) = p_2(\xi)k_{21}M'_1 \,, \qquad (77)$$

where $p_\alpha(\xi)$ represents the probability that an α-th type active center positioned on the end of a terminal block of length ξ resides in phase β close to its boundary. Since the time scale of the addition of a monomer to the growing macroradical as a rule appreciably exceeds the time scale of attaining by this macroradical conformational equilibrium, the dependence $p_\alpha(\xi)$ has to be found from an equilibrium theory.

In the framework of this theory, the probability $p_\alpha(\xi)$ is equal to the product of two factors. The first of them, $q_\alpha(\xi)$, equals the probability for the growing end of an α-th type block to be situated in the neighborhood of the interphase boundary on the α-th phase side. The second factor, $\kappa_\alpha = \exp\{-\Delta F_{\alpha\beta}/k_B T\}$, is controlled by the loss in free energy $\Delta F_{\alpha\beta}$ of a α-th type macroradical under the transition of its terminal unit through the interface from phase α to phase β. The more is the ratio of this quantity to the product of the Boltzmann constant k_B and temperature T, the less the probability for the terminal unit of the macroradical to cross the interface. The problem of finding the dependence $q_\alpha(\xi)$ for the Gaussian polymer chain reduces to the consideration of the random walks in half-space over the plane with a reflecting boundary. Under such a consideration, the unknown function $q_\alpha(\xi)$ will equal the probability to find among all trajectories with length ξ such a trajectory where both ends are located on the plane. The solution of this problem yields at $\xi \gg 1$ asymptotic dependence $q_\alpha(\xi) = c\xi^{-1/2}$ where c is the numerical coefficient of the order of unity [77]. As a result, the following expression for function $w_\alpha(\eta)$ (Eq. 75) is arrived at

$$w_\alpha(\eta) = \exp\left\{- 2\varepsilon_\alpha \eta^{1/2} - \varepsilon_\alpha^{(0)}\eta\right\} \qquad (\alpha \neq \beta = 1, 2)\,, \qquad (78)$$

where the following designations are used

$$\varepsilon_\alpha = \frac{cM'_\beta \kappa_\alpha}{2M'_\alpha r_\alpha} \qquad \varepsilon_\alpha^{(0)} = \frac{T_\alpha}{\theta_\alpha} = \frac{k_{t\alpha}R'_\alpha}{k_{\alpha\alpha}M'_\alpha} \qquad (79)$$

The values of the reactivity ratios $r_1 = k_{11}/k_{12}$ and $r_2 = k_{22}/k_{21}$ involved in the first of formulas (Eq. 79) are available in the literature for hundreds of monomeric pairs which are employed in radical copolymerization [78].

Substituting expressions (Eq. 78) for $w_1(\eta)$ and $w_2(\eta)$ into relationships Eqs. 73 and 74 we get a closed set of kinetic equations describing radical copolymerization in the framework of the simplest model in hand. The values of the rates of initiation in phases 1 and 2 entering in Eq. 74 are determined as follows

$$I_1 = k_1^{\text{in}} \Theta s \qquad I_2 = k_2^{\text{in}} \Theta s\,, \tag{80}$$

where Θ denotes the surface concentration of an initiator molecule. Here the rate constant of the initiation reaction in the α-th phase, k_α^{in}, is proportional to the probability that the decomposition of the labile bond of the initiator molecule will happen just in the α-th phase as well as to the rate constant of this bond decomposition and the efficiency of a polymer chain initiation in the α-th phase.

It should be emphasized that from the standpoint of the statistical chemistry of macromolecular reactions the mathematical model of the interphase copolymerization in point substantially differs from other models describing the synthesis of copolymers. Centrally important among such distinctions are the two below.

Firstly, the distribution of blocks for lengths is governed not only by the rate constants of the propagation reactions and monomers' concentrations, but also by conformations of the growing polymer chain. That is the interphase copolymerization is a prominent example of conformationally dependent design of macromolecules.

Secondly, the stochastic process of the conventional movement along copolymer chain is non-Markovian, and, which is more essential, cannot be reduced to this by any "coloring" of monomeric units [1]. This conclusion follows immediately from the fact that the infinitesimal probability $V_{\alpha\beta}$ (Eq. 77) of the transition from state S_α into another state S_β depends on the residence time of the realizations of a stochastic process in the initial state. Otherwise stated, this stochastic process possesses an infinitely long memory, and consequently, is irreducible to the Markovian stochastic processes by any extension of the set of its states. Such a reduction might be possible, provided the probability $V_{\alpha\beta}$ had been independent of the states preceding the initial one. In this case, integral Eqs. 73 and 74 describing the stochastic process with infinitely long memory, reduce to differential equations (Eq. 76) describing a Markovian stochastic process.

The major problem challenging a quantitative theory of a copolymerization is the derivation of the expressions for the rate of this process and for the statistical characteristics of the chemical structure of its products. Among the latter in the case of multiblock copolymers is the size-composition distribution of macromolecules, their distribution for numbers of blocks, as well as the distribution of these blocks for their length.

5.4 Kinetics

The overall rate V of the interphase copolymerization equals the sum of rates of the formation of homopolymer blocks in each phase

$$V = -\frac{\mathrm{d}M}{\mathrm{d}t} = V_1 + V_2 = \theta_1 R_1 + \theta_2 R_2 . \tag{81}$$

Here R_1 and R_2 stand for concentrations of the active centers of types 1 and 2

$$R_\alpha \equiv \int_0^\infty R_\alpha(l)\mathrm{d}l \qquad (\alpha = 1, 2) , \tag{82}$$

while concentrations of radicals $R_\alpha(l)$ (Eq. 73) are found from Eq. 74. Resorting to the Laplace transformation, it is possible to obtain their solution and get exact analytical expressions for R_1 and R_2

$$R_1 = \frac{(1 - H_1)(I_1 + I_2 H_2)}{T_1(1 - H_1 H_2)} ; \qquad R_2 = \frac{(1 - H_2)(I_1 H_1 + I_2)}{T_2(1 - H_1 H_2)} . \tag{83}$$

In these formulas the following designations are used

$$H_\alpha = H(b_\alpha) , \quad H(b) = \sqrt{\pi b}\exp(b)\mathrm{erfc}(\sqrt{b}) , \quad b_\alpha = \varepsilon_\alpha^2 / \varepsilon_\alpha^{(0)} , \tag{84}$$

where the symbol "erfc" denotes the function, referred to as "addition to the probability integral" [79], whereas parameters ε_α and $\varepsilon_\alpha^{(0)}$ have been defined above (Eq. 79).

Interestingly enough, quantity H_α (Eq. 84) has a rather transparent probabilistic meaning. In fact, the growth of the terminal α-th type block of a macroradical may be over either by the transition of an active center into another phase, or by its vanishing due to the chain termination reaction. The probabilities of these events, coinciding with the probabilities that a block chosen at random will be either internal or external, are equal to H_α and $1 - H_\alpha$, respectively.

Depending on values of parameters b_1 and b_2, it is possible to distinguish three limiting regimes of the interphase copolymerization

$$1)\ b_1 \gg 1 , \quad b_2 \gg 1 \quad 2)\ b_1 \ll 1 , \quad b_2 \ll 1 \quad 3)\ b_1 \gg 1 , \quad b_2 \gg 1 . \tag{85}$$

In the first of these limiting regimes both quantities, $H_1 = H(b_1)$ and $H_2 = H(b_2)$, turn out to be close to unity. This means that the fraction of the internal blocks substantially exceeds the fraction of the external ones, i.e. copolymerization products under this regime are multiblock copolymers. A completely different type of situation occurs in the limiting regime 2 where quantities, H_1 and H_2, are much less than unity. In this regime, almost all blocks are external and, consequently, a mixture of molecules of two homopolymers is formed in the course of copolymerization. As for the third of the regimes (Eq. 85), here the value of quantity H_1 is very close to unity, while H_2 is small enough. Under this regime the copolymerization products will be homopolymer molecules with the second type units and those of the diblock copolymer being formed under the initiation of chains in the second and the first phases, respectively. Obviously, the most promising from a practical viewpoint is the regime 1 that will be addressed in more detail below.

In this regime expressions (Eq. 83) for finding the concentrations of active centers are

$$R_1 = \frac{Ib_1^{-1}}{T_1(b_1^{-1}+b_2^{-1})}; \quad R_2 = \frac{Ib_2^{-1}}{T_2(b_1^{-1}+b_2^{-1})}, \text{ where } I = I_1 + I_2 . \tag{86}$$

Substituting their solutions into formulas (Eq. 82), we will get the expression for the rate of monomer M_α depletion

$$V_\alpha = \frac{1}{\varepsilon_\alpha^2}\sqrt{\frac{Is}{\Lambda}} \quad (\alpha = 1,2), \text{ where } \Lambda = \frac{k_{t1}}{d_1\theta_1^2\varepsilon_1^4} + \frac{k_{t2}}{d_2\theta_2^2\varepsilon_2^4} \tag{87}$$

that leads in the simplest case $d_1 = d_2 = d$ to the following formula for the interphase copolymerization rate

$$V = (Isd)^{1/2}\left[\rho_1^2(M_1')^4 + \rho_2^2(M_2')^4\right]\left[\delta_1^2\rho_1^4(M_1')^6 + \delta_2^2\rho_2^4(M_2')^6\right]^{-1/2} . \tag{88}$$

Here $\rho_\alpha = r_\alpha/\kappa_\alpha$, while $\delta_\alpha = k_{t\alpha}^{1/2}/k_{\alpha\alpha}$ represents the parameter entering in the traditional expression for homopolymerization rate [76]

$$V_\alpha^{\text{hom}} = \sqrt{I}M_\alpha/\delta_\alpha . \tag{89}$$

The values of this parameter are extensively reported in the literature for many monomers [80]. Using expressions (Eq. 80), it is easy to note that the overall number of moles of monomers being polymerized in unit time in a reaction system is proportional to the interphase surface of the miniemulsion.

To have expression (Eq. 88) compared with that derived for the rate of the traditional homophase copolymerization

$$V = I^{1/2}\left(r_1M_1^2 + r_2M_2^2\right)\left(\delta_1^2r_1^2M_1^2 + \delta_2^2r_2^2M_2^2\right)^{-1/2} \tag{90}$$

recourse should be made to relationship $M_\alpha' = M_\alpha/y^\alpha$, connecting concentrations of monomer M_α in the α-th phase, M_α', and throughout the reaction system, M_α. The comparison indicates that the dependencies of the rate V of the depletion of monomers on their overall concentration $M = M_1 + M_2$ and on the initiation rate I are the same for homophase and interphase copolymerization. However, the dependence of V on monomer mixture composition $\boldsymbol{x}$ is qualitatively different for these two processes.

5.5 Statistical Characteristics of the Chemical Structure of Macromolecules

An exhaustive theoretical characterization of the chemical structure of polydisperse block copolymers suggests knowledge of the distributions of:

1) blocks for length (Eq. 4);
2) macromolecules for numbers of blocks;
3) macromolecules for size and composition (Eq. 1).

Expressions for these three distributions for the products of the interphase copolymerization are presented in paper [74]. Below we will discuss briefly these results.

Under theoretical examination of macromolecules of multiblock copolymers, it is necessary to distinguish external blocks located at the ends of macromolecules from the internal ones. Expressions for their distributions for length η

$$f_\alpha^{\mathrm{bl(ex)}}(\eta) = \frac{\varepsilon_\alpha^{(0)} w_\alpha(\eta)}{1 - H_\alpha} \qquad f_\alpha^{\mathrm{bl(in)}}(\eta) = \frac{w_\alpha(\eta)\varepsilon_\alpha}{\sqrt{\eta} H_\alpha} \tag{91}$$

are readily derived from simple probabilistic reasoning that relies upon the formalism of the Renewal Theory [75]. The appearance of function $w_\alpha(\eta)$ (Eq. 78) indicates that the distributions (Eq. 91) markedly differ from exponential distribution (Eq. 12), peculiar to the Markovian copolymers.

As it ensues from asymptotic expressions

$$\overline{\eta}_\alpha^{\mathrm{in}} \equiv \int_0^\infty \eta f_\alpha^{\mathrm{bl(in)}}(\eta)\mathrm{d}\eta = \frac{1}{2\varepsilon_\alpha^2} = \pi \left[\frac{x_\alpha(1 - y^\alpha)r_\alpha}{(1 - x_\alpha)y^\alpha \kappa_\alpha}\right]^2 \qquad \overline{\eta}_\alpha^{\mathrm{ex}} = 3\overline{\eta}_\alpha^{\mathrm{in}} \tag{92}$$

average lengths of the internal and external blocks of α-th type units increase with the growth of molar fraction x_α of monomer M_α in the reaction system and reactivity ratio r_α. Both of these factors, enlarging the probability of the addition to the propagating radical R_α of monomer M_α rather than monomer M_β, are inherent in homophase copolymerization. However, in parallel with two aforementioned factors, there is one more peculiar to the interphase mechanism of the copolymer synthesis responsible for the formation of long blocks. This factor is minor thermodynamic affinity between hydrophilic and hydrophobic monomers as well as between their units in macromolecules. The less pronounced is this affinity the smaller is the thermodynamic parameter κ_α and, thus, the higher, according to formula (Eq. 92), the average lengths of blocks of the corresponding type in copolymer chains.

The problem of finding the joint distribution of macromolecules $P(n_1, n_2; m_1, m_2)$ for numbers of internal, (n_1, n_2), and external, (m_1, m_2), blocks can be easily solved by the statistical method [1, 3]. This is possible because the succession of blocks in a macromolecule is described by the absorbing Markov chain with the transition matrix $\boldsymbol{Q}^{ab}$ and the vector of initial states $\boldsymbol{\nu}$ whose components ν_1, ν_2

$$\boldsymbol{Q}^{ab} = \begin{pmatrix} 1 & 0 & 0 \\ \nu_{10} & 0 & \nu_{12} \\ \nu_{20} & \nu_{21} & 0 \end{pmatrix} \qquad \begin{array}{l} \nu_1 = I_1/I, \quad \nu_2 = I_2/I \\ \nu_{12} = 1 - \nu_{10} = H_1 \\ \nu_{21} = 1 - \nu_{20} = H_2 \end{array} , \tag{93}$$

where quantities H_α have been defined above (Eq. 84). Having invoked the mathematical apparatus of the theory of the Markov chains, authors of paper [74] derived an exact formula for the generating function of distribution $P(n_1, n_2; m_1, m_2)$, which permits finding its explicit analytical appearance. They established, particularly, that in macromolecules synthesized at the interphase boundary in regime 1 (Eq. 85), distributions $P(n_1)$ and $P(n_2)$ asymptotically coincide being described by an exponential function. Average values of numbers of different blocks in this regime

$$\overline{n}_1 = \overline{n}_2 = \frac{2b_1 b_2}{b_1 + b_2}, \qquad \overline{m}_1 = \frac{b_2}{b_1 + b_2} \qquad \overline{m}_2 = \frac{b_1}{b_1 + b_2} \tag{94}$$

are controlled only by parameters b_1 and b_2 (Eq. 84). If one of them is appreciably less than the other, just this parameter will predetermine the average number of blocks contained in a macromolecule.

An interesting approach has been employed in paper [74] to find the distribution $f(l_1, l_2)$ of copolymer chains for numbers l_1 and l_2 of monomeric units $\overline{M}_1$ and $\overline{M}_2$. This distribution is evidently equivalent to the SCD, because the pair of numbers l_1 and l_2 unambiguously characterizes chemical size ($l = l_1 + l_2$) and composition ($\xi_1 = l_1/l, \xi_2 = l_2/l$) of a macromolecule. The essence of this approach consists of invoking the Superposition Principle [81] that enables the problem of finding the Laplace transform $G(p_1, p_2)$ of distribution $f(l_1, l_2)$ to be reduced to the solution of two subsidiary problems. The first implies the derivation of the expression for the generating function $U(z_1^{in}, z_2^{in}; z_1^{ex}, z_2^{ex})$ of distribution $P(n_1, n_2; m_1, m_2)$, and the second is concerned with finding the Laplace transforms $g_\alpha^{in}(p_1, p_2)$ and $g_\alpha^{ex}(p_1, p_2)$ of distributions (Eq. 91). With these two problems solved, it is possible to obtain the characteristic function $G(p_1, p_2)$ of distribution $f(l_1, l_2)$ using the Superposition Principle formula

$$G(p_1, p_2) = U\left(g_1^{in}, g_2^{in}; g_1^{ex}, g_2^{ex}\right) . \tag{95}$$

The recourse to the above-described procedure permitted the derivation [74] of an exact expression for function $G(p_1, p_2)$. It is of utmost importance for the construction of a phase diagram of a melt or solution of interphase copolymerization products, since this function enters in the equations for the cloud point curve [82].

Besides, the statistical moments of the SCD may be found by differentiating function (Eq. 95) with respect to its arguments p_1 and p_2. For instance, the average number of α-th type units in a macromolecule is calculated by formula

$$\bar{l}_\alpha = -\left.\frac{\partial G_N}{\partial p_\alpha}\right|_{p_1=p_2=0} = \overline{n_\alpha \eta_\alpha^{\text{in}}} + \overline{m_\alpha \eta_\alpha^{\text{ex}}} . \tag{96}$$

Using relationships Eqs. 92 and 94, it is an easy matter to find average values of chemical size $\bar{l} = \bar{l}_1 + \bar{l}_2$ and composition $X_\alpha = \bar{l}_\alpha / \bar{l}$ of a copolymer obtained in regime 1 (Eq. 85). The expressions for them in terms of the parameters (Eq. 79) have a simple form

$$\bar{l} = \frac{1}{\varepsilon_1^{(0)} X_1 + \varepsilon_2^{(0)} X_2} \qquad X_\alpha = \frac{\varepsilon_\alpha^{-2}}{\varepsilon_1^{-2} + \varepsilon_2^{-2}} . \tag{97}$$

In this regime, macromolecules of the ergodic copolymer [1, 85] are formed whose SCD has exactly the same appearance as in the case of traditional free-radical copolymerization [83, 84]. Such SCD (Eq. 1) represents the product of two functions. The first is the distribution of macromolecules for their size l, whereas the second is fractional distribution $W(l|\zeta)$ of macromolecules with a fixed number of units l for their composition $\zeta \equiv \zeta_1 = l_1/l$. Function $W(l|\zeta)$ is described by the Gauss formula with the average value $\bar{\zeta} = X_1$ (Eq. 97) and dispersion $\sigma_l^2 = D/l$. Parameter D as well as X_1 is independent of l and may be found by the methods of the statistical chemistry of polymers [85]. However, for real polymer chains with large values of l, the dispersion σ_l^2 of the distribution $W(l|\zeta)$, being reciprocal to l, is so small that the composition inhomogeneity of macromolecules formed at fixed monomer mixture composition may be neglected. It is a very good approximation because this *instantaneous* composition inhomogeneity is normally far less pronounced as compared to the *conversional* composition inhomogeneity originated by monomer composition drift in the course of copolymerization. Only at its initial stages when such an evolution can be ignored, one may use the above-presented theoretical formulas. In order to extend the quantitative theory of interphase copolymerization to the whole range of the monomers' conversions, one should consider this processes dynamics just in the same way as under the theoretical description of the traditional homophase copolymerization [84].

5.6 Conversional Evolution of a Copolymer Composition Distribution

Under the quasi-homogeneous approach the monomer mixture composition is characterized by vector $\boldsymbol{x}$ with components $x_1 = M_1/M$ and $x_2 = M_2/M$, whose drift with conversion is described by equations [84]

$$(1-p)\frac{\mathrm{d}x_\alpha}{\mathrm{d}p} = x_\alpha - X_\alpha(\boldsymbol{x}) , \qquad x_\alpha(0) = x_\alpha^0 \qquad (\alpha = 1, 2) , \tag{98}$$

where $X_\alpha(\boldsymbol{x})$ represents the molar fraction of the α-th type units in a copolymer macromolecule being formed at the monomer composition $\boldsymbol{x}$. Components $X_1(\boldsymbol{x})$, $X_2(\boldsymbol{x})$ (Eq. 97) of vector $\boldsymbol{X}(\boldsymbol{x})$ of the *instantaneous* composition of a copolymer, which is formed under regime 1 (Eq. 85) at fixed $\boldsymbol{x}$ in the proximity of the interface, can be calculated using the following expression

$$X_\alpha(\boldsymbol{x}) \equiv \frac{dM_\alpha}{dM} \equiv \frac{V_\alpha}{V} = \frac{\varepsilon_\alpha^{-2}}{\varepsilon_1^{-2} + \varepsilon_2^{-2}} = \frac{\hat{\rho}_\alpha^2 x_\alpha^4}{\hat{\rho}_1^2 x_1^4 + \hat{\rho}_2^2 x_2^4}, \tag{99}$$

$$\text{where } \hat{\rho}_\alpha = \frac{\rho_\alpha}{(y_\alpha)^2}, \quad \rho_\alpha = \frac{r_\alpha}{\kappa_\alpha}.$$

By virtue of the conditions $x_1 + x_2 = 1$, $X_1 + X_2 = 1$, only one of two equations (Eq. 98) (e.g. the first one) is independent. Analytical integration of this equation results in explicit expression connecting monomer composition $\boldsymbol{x}$ with conversion p. This expression in conjunction with formula (Eq. 99) describes the dependence of the instantaneous copolymer composition $\boldsymbol{X}$ on conversion. The analysis of the results achieved revealed [74] that the mode of the drift with conversion of compositions $\boldsymbol{x}$ and $\boldsymbol{X}$ differs from that occurring in the processes of homophase copolymerization. It was found that at any values of parameters $\hat{\rho}_1$, $\hat{\rho}_2$ and initial monomer composition $\boldsymbol{x}^0$ both vectors, $\boldsymbol{x}$ and $\boldsymbol{X}$, will tend with the growth of p to common limit $\boldsymbol{x}^* = \boldsymbol{X}^*$. In traditional copolymerization, systems also exist in which the instantaneous composition of a copolymer coincides with that of the monomer mixture. Such a composition, $\boldsymbol{x}^* = \boldsymbol{X}^*$, is known as the "azeotrop". Its values, controlled by parameters of the model, are defined for homophase (a) [1, 86] and interphase (b) copolymerization as follows

$$\text{a) } x_1^* = \frac{1 - r_2}{2 - r_1 - r_2}, \qquad x_2^* = \frac{1 - r_1}{2 - r_1 - r_2} \tag{100}$$

$$\text{b) } x_1^* = \frac{\hat{\rho}_2^{2/3}}{\hat{\rho}_1^{2/3} + \hat{\rho}_2^{2/3}} \equiv a_2, \quad x_2^* = \frac{\hat{\rho}_1^{2/3}}{\hat{\rho}_1^{2/3} + \hat{\rho}_2^{2/3}} \equiv a_1.$$

Under homophase synthesis in real systems the azeotrop (a) exists only provided $r_1 < 1$ and $r_2 < 1$. In this case, however, it is a repeller, unlike in the case of interphase copolymerization where the azeotrop (b) is an attractor. This means that at the final stage of homophase copolymerization homopolymer molecules are primarily formed in all real systems whereas under the interphase synthesis the majority of copolymer chains formed at $p \to 1$ have the azeotropic composition $\boldsymbol{x}^*$.

Given the dependence of $\boldsymbol{X}$ on p, it is possible resorting to the general algorithm [84] to derive an expression for the fraction of monomeric units $f^c(\zeta; p)$ (Eq. 1) involved in molecules with composition ζ, which are formed during the interphase copolymerization under all conversions p' preceding p

$$f^{c}(\boldsymbol{\zeta};p) = \langle \delta(\boldsymbol{\zeta} - \boldsymbol{X}) \rangle \equiv \frac{1}{p}\int_{0}^{p} \delta(\boldsymbol{\zeta} - \boldsymbol{X}(p'))\mathrm{d}p' \tag{101}$$

$$= \frac{1}{p}\frac{\left[(a_1\zeta_1)^{1/4} - (a_2\zeta_2)^{1/4}\right]^{-2/3}}{h(\boldsymbol{x}^0)4(a_1a_2)^{1/4}(\zeta_1\zeta_2)^{5/4}}\left[\frac{(a_2^3\zeta_1)^{1/4} + (a_1^3\zeta_2)^{1/4}}{(a_1\zeta_1)^{1/2} + (a_2\zeta_2)^{1/2} + (a_1a_2\zeta_1\zeta_2)^{1/4}}\right]^{4/3}$$

When deriving this expression for the average composition distribution, authors of paper [74] entirely neglected its instantaneous constituent, having taken (as is customary in the quantitative theory of radical copolymerization [3, 84]) the Dirac delta-function $\delta(\boldsymbol{\zeta} - \boldsymbol{X})$ as the *instantaneous* composition distribution. Its averaging over conversions, denoted hereinafter by angular brackets, leads to formula (Eq. 101). Note, this formula describes the composition distribution only provided copolymer composition $\boldsymbol{\zeta}$ falls in the interval between $\boldsymbol{X}(0)$ and $\boldsymbol{X}(p)$. Otherwise, this distribution function vanishes at all values of composition $\boldsymbol{\zeta}$ lying outside the above-mentioned interval.

The form of the distribution (Eq. 101), as shown in Fig. 7, qualitatively differs from that exhibited by this distribution for the products of homophase copolymerization. This distinction takes place both in real systems ($r_1 < 1, r_2 < 1$), where statistical copolymers are formed and in hypothetical systems ($r_1 \gg 1, r_2 \gg 1$), where the formation of multiblock copolymers is expected. Essentially, the composition distribution in the latter systems

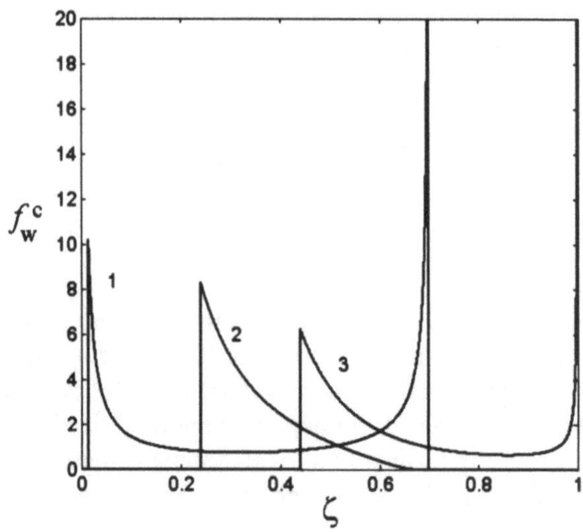

Fig. 7 Appearance of the composition distribution function $f^{c}(\zeta; 1)$ typical of the products of interphase (1) as well as hypothetical (2) and real (3) homophase copolymerization at $p = 1$

lacks a mode whose maximum corresponds to macromolecules of azeotropic composition. This mode, however, is present in the composition distribution of the interphase copolymerization products. As can be seen from Fig. 7, the composition distribution of the final products of traditional homophase copolymerization can be bimodal as well. However, this distribution (curve 3) qualitatively differs from that of composition distribution of the products of interphase copolymerization (curve 1) by the position of the mode with infinite height. In the first case, this mode corresponds to a homopolymer whereas in the second case it corresponds to an azeotropic block copolymer. As for another mode, whose amplitude is finite, it for all curves depicted in Fig. 7 corresponds to a copolymer formed in the very beginning of the process. This will be either a multiblock copolymer (curves 1 and 2) or a statistical copolymer (curve 3).

Among important characteristics of composition distribution (Eq. 101) are its statistical moments of the first and the second order

$$\int_0^1 \zeta_1 f^c(\zeta;p)\mathrm{d}\zeta_1 = \langle X_1 \rangle = \frac{1}{p}\int_0^p X_1(p')\mathrm{d}p' = \frac{1}{p}\left[x_1^0 - (1-p)x_1\right] \tag{102}$$

$$\int_0^1 (\zeta_1 - \langle X_1 \rangle)^2 f^c(\zeta;p)\mathrm{d}\zeta_1 \equiv \sigma^2(p) = \frac{1}{p}\int_0^p X_1^2(p')\mathrm{d}p' - \langle X_1 \rangle^2 . \tag{103}$$

Formulas (Eqs. 102 and 103) enable one to calculate readily the dependence on conversion of the principal statistical characteristics of composition distribution (Eq. 101). The results of such calculations at a fixed value of parameter $a_1 = 1 - a_2$ (Eq. 100) are depicted in Fig. 8.

Figure 8a demonstrates that the monomer mixture composition x either increases or decreases with conversion p, provided its initial value x^0 is either less or more than azeotropic composition $x^* = 0,7$. An analogous behavior is demonstrated with increasing p by instantaneous X (Fig. 8b) and average $\langle X \rangle$ (Fig. 8c) copolymer compositions. Both curves characterizing these compositions have value $X(x^0)$ as their starting point. However, the first ends at azeotrop $X^* = x^*$ whereas the second ends at point x^0. Figure 8d presents the dispersion σ^2 (Eq. 103) of the composition distribution that is a qualitative characteristic of composition inhomogeneity of copolymerization products. This dispersion monotonically increases with conversion, changing from $\sigma^2(0) = 0$ up to $\sigma^2(1) = \sigma^2_{max}$. The quantity σ^2_{max} is essentially controlled by the initial composition of monomers x^0. It vanishes at points $x^0 = 0$, $x^0 = 1$ and $x^0 = x^* = 0,7$, taking on the larger value the more x^0 is spaced from these three points. As follows from Fig. 8d for the majority of the initial compositions x^0 the value of σ^2_{max} lies within the range 10^{-2}–10^{-1}. This value is one order of magnitude larger than that which is typical of homophase copolymerization [80]. The aforementioned conclusions are of

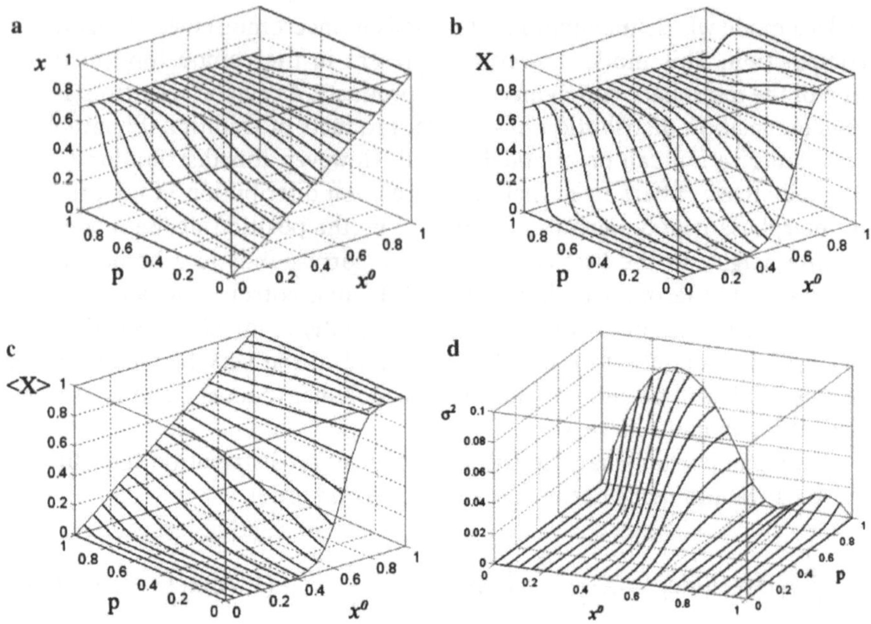

Fig. 8 Dependencies on conversion p of monomer mixture composition x (**a**), instantaneous X (**b**) and average $\langle X \rangle$ (**c**) copolymer composition as well as dispersion σ^2 (**d**) of the composition distribution calculated at different values of the initial compositions of monomers x^0. The calculations have been carried out at values of parameters a_1 and $a_2 = 1 - a_1$ (Eq. 100) equal to 0.3 and 0.7, respectively

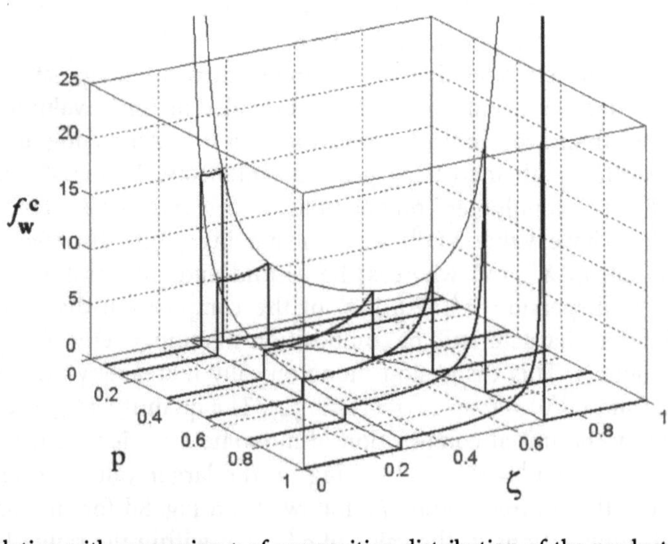

Fig. 9 Evolution with conversion p of composition distribution of the products of interphase copolymerization calculated at the initial monomer mixture composition $x^0 = 0.6$ and parameter a_1 (Eq. 100) equal to 0.3

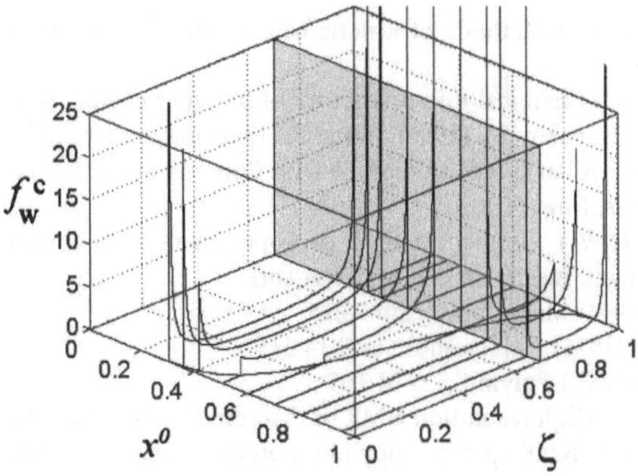

Fig. 10 Dependence of the composition distribution of a copolymer formed under complete conversion of monomers on their initial composition x^0. The diagrams are presented here for interphase copolymerization when $a_1 = 0.3$

general significance since the qualitative appearance of the curves presented in Fig. 8 remains the same at arbitrary values of parameter a_1.

Of considerable theoretical and practical interest is the answer to the question how the composition distribution of the products of interphase copolymerization changes throughout this process. One may get an idea about the peculiarities of such a change by turning to Fig. 9. The inspection of the curves presented shows how with the rise of conversion the broadening of the composition distribution occurs. This is accompanied by simultaneous formation of its mode whose maximum corresponds at a given conversion to a copolymer composition, which is the closest to the azeotropic one. Under complete conversion ($p = 1$), such a maximum coincides with azeotrop $x^* = X^*$ and has an infinite height. It can be readily seen in Fig. 10 that with a decreased fraction of any monomer in the initial mixture, the second mode with the maximum corresponding to the initial copolymer composition emerges on the curve of the composition distribution of the copolymerization final products. The height of this maximum grows to diverge at point $x^0 = 0$ or $x^0 = 1$.

Acknowledgements Support of this work by the Alexander-von-Humboldt Foundation (Program for investment in the Future (ZIP)) and INTAS (Project 01-607) is gratefully acknowledged.

References

1. Kuchanov SI (2000) Adv Polym Sci 152:157
2. Lowry GG (ed) (1989) Markov chains and Monte Carlo calculations in polymer science. Marcel Dekker Inc, New York

3. Kuchanov SI (1978) Methods of kinetic calculations in polymer chemistry. Khimia Publ, Moscow
4. Koenig JL (1980) Chemical microstructure of polymer chains. Wiley, New York
5. Kuchanov SI, Panyukov SV (1996) Statistical thermodynamics of heteropolymers and their blends. In: Allen G (ed) Comprehensive Polymer Science, second suppl, chap 13. Pergamon Press, New York
6. Glockner G (1991) Gradient HPLC of copolymers and chromatographic cross-fraction. Springer, Berlin Heidelberg New York
7. Cowie JMG (1994) Macromol Symp 78:15
8. Johnston NW (1976) J Macromol Sci C 14:215
9. Furukawa J (1975) J Polym Sci C 51:105
10. Bovey FA (1972) High resolution NMR of macromolecules. Academic Press, New York
11. Tonelli AE (1989) NMR spectroscopy and polymer microstructure. Wiley, New York
12. Katime IA, Quintana JR (1989) In: Allen G (ed) Comprehensive polymer science, vol 1. Pergamon Press, Oxford, p 103
13. Benoit H, Froelich D (1972) In: Huglin MB (ed) Light scattering from polymer solutions. Pergamon Press, London, p 467
14. Aliev M, Kuchanov S (2005) Eur Phys J B 43:251
15. Binder K (1994) Adv Polym Sci 112:181
16. Plate NA, Litmanovich AD, Noa OV (1995) Macromolecular reactions. Wiley, Chichester, UK
17. Kuchanov SI (1996) In: Kuchanov SI (ed) Mathematical methods in contemporary chemistry. Gordon & Breach, Amsterdam, p 267
18. Litmanovich AD (1978) Dokl Acad Nauk SSSR 240:111
19. Litmanovich AD (1980) Eur Phys J 16:269
20. Govorun EN, Ivanov VA, Khokhlov AR, Khalatur PG, Borovinsky AL, Grosberg AYu (2001) Phys Rev E 64:40903
21. Paul W, Baschnagel J (1999) Stochastic processes, chap 4. Springer, Berlin Heidelberg New York
22. Kuchanov SI, Khokhlov AR (2003) J Chem Phys 118:4672
23. Kuchanov SI, Zharnikov TV, Khokhlov AR (2003) Eur Phys J E 10:93
24. Aris R (1965) Introduction to the analysis of chemical reactors. Prentice Hall, Englewood Cliffs, New Jersey
25. Astarita G (1967) Mass transfer with chemical reaction. Elsevier, Amsterdam
26. Khokhlov AR, Khalatur PG (1998) Physica A 249:253
27. Khokhlov AR, Khalatur PG (1999) Phys Rev Lett 82:3456
28. Zheligovskaya EA, Khalatur PG, Khokhlov AR (1999) Phys Rev E 59:3071
29. Ivanov VA, Chertovich AV, Lazutin AA, Shusharina NP, Khalatur PG, Khokhlov AR (1999) Macromol Symp 146:259
30. Lifshitz IM, Grosberg AYu, Khokhlov AR (1978) Rev Mod Phys 50:683
31. Higgins JS, Benoit H (1974) Polymers and neutron scattering. Oxford Science, New York
32. Hamley IW (1998) The physics of block copolymers. Oxford University Press, Oxford
33. Bates FS, Fredrickson GH (1990) Ann Rev Phys Chem 41:525
34. Binder K (1994) Adv Polym Sci 112:181
35. Toledano J-C, Toledano P (1987) The Landau theory of phase transitions. World Scientific, Singapore
36. Shakhnovich EI, Gutin AM (1989) J Phys (France) 50:1843
37. Panyukov SV, Kuchanov SI (1991) JETP Lett 54:501
38. Panyukov SV, Kuchanov SI (1992) J Phys II (France) 2:1973

39. Kuchanov SI, Zharnikov TV, Khokhlov AR (2006) Phase behaviour of proteinlike copolymers obtained by polymeranalogous transformation of homopolymer globule (paper to be published)
40. Kuchanov SI, Panyukov SV (2006) A new look at the Landau theory of phase transitions in polydisperse heteropolymer liquids (paper to be published)
41. Frensdorff HK, Pariser R (1963) J Chem Phys 39:2303
42. Greenley ZR (1980) J Macromol Sci A 14:445
43. Mayo FR, Lewis FM (1944) J Am Chem Soc 66:1594
44. Alfrey T, Goldfinger G (1944) J Chem Phys 12:205
45. Merz E, Alfrey T, Goldfinger G (1946) J Polym Sci 1:75
46. Ham GE (1960) J Polym Sci 45:169
47. Gaylord NG, Takahashi A (1969) Adv Chem ACS # 91:94
48. Seiner JA, Litt M (1971) Macromolecules 4:308
49. Hill DJ, O'Donnell JJ, O'Sullivan PW (1982) Progr Polym Sci 8:215
50. Ratzsch M, Vogl O (1991) Progr Polym Sci 16:279
51. Kuchanov SI, Russo S (1997) Macromolecules 30:4511
52. Plochocka K, Harwood HJ (1978) ACS Polym Preprints 19:240
53. Harwood HJ (1987) Makromol Chem, Macromol Symp 10:331
54. Toppet S, Slinckx M, Smets G (1975) J Polym Sci A 13:1879
55. Park KY, Santee ER, Harwood HJ (1986) ACS Polym Preprints 27:240
56. Suggate JR (1978) Makromol Chem 179:1219
57. Suggate JR (1979) Makromol Chem 180:679
58. Park KY, Santee ER, Harwood HJ (1989) Eur Polym J 25:651
59. Davis TP (1990) Polym Commun 31:442
60. Klumperman B, Brown PG (1994) Macromolecules 27:6100
61. Semchikov YuD, Knjazeva TE, Smirnova LA, Bazhenova NN, Slavnitskaya NN (1981) Vysokomol Soedin B 23:483
62. Smirnova LA, Semchikov YuD, Slavnitskaya NN, Knjazeva TE, Modeva ShI, Bulgakova SA (1982) Dokl Acad Nauk USSR 263:1170
63. Semchikov YuD, Smirnova LA, Knjazeva TE, Bulgakova SA, Voskoboynik GA, Sherstyanykh VI (1984) Vysokomol Soedin A 26:704
64. Semchikov YD, Smirnova LA, Bulgakova SA, Sherstyanykh VI, Knjazeva TE, Slavnitskaya NN (1987) Vysokomol Soedin B 29:220
65. Semchikov YuD, Knjazeva TE, Smirnova LA, Riabov SA, Slavnitskaya NN, Modeva ShI, Bulgakova SA, Sherstyanykh VI (1987) Vysokomol Soedin A 29:2625
66. Semchikov YuD, Smirnova LA, Bulgakova SA, Knjazeva TE, Sherstyanykh VI, Slavnitskaya NN (1988) Dokl Acad Nauk USSR 298:411
67. Egorochkin GA, Semchikov YuD, Smirnova LA, Knjazeva TE, Tikhonova ZA, Kariakin NV, Sveshnikova TG (1989) Vysokomol Soedin B 31:46
68. Semchikov YuD, Smirnova LA, Sherstyanykh VI (1989) Vysokomol Soedin B 31:249
69. Egorochkin GA, Smirnova LA, Semchikov YuD, Tikhonova ZA (1989) Vysokomol Soedin B 31:433
70. Smirnova LA, Egorochkin GA, Semchikov YuD, Kariakin NV (1990) Dokl Acad Nauk USSR 313:381
71. Smirnova LA, Semchikov YuD, Egorochkin GA, Sveshnikova TG, Konkina TN (1990) Vysokomol Soedin B 32:624
72. Semchikov YuD, Smirnova LA, Knjazeva TE, Sherstyanykh VI (1990) Eur Polym J 26:883
73. Semchikov YuD, Slavnitskaya NN, Smirnova LA, Sherstyanykh VI, Sveshnikova TG, Borina TI (1990) Eur Polym J 26:889

74. Smirnova LA, Semchikov YuD, Kopylova NA, Sveshnikova TG, Izvolenskii VV, Grebneva MV (1991) Dokl Acad Nauk USSR 317:410
75. Semchikov YuD, Smirnova LA, Kopylova NA, Sveshnikova TG (1994) Vysokomol Soedin B 37:542
76. Smirnova LA (1991) PhD thesis, Moscow State University
77. Dondos A, Benoit H (1976) In: Olc K (ed) Order in polymer solutions. Gordon & Breach Sci Publ, London, p 175
78. Katime J, Garro P, Teijon JM (1976) Eur Polym J 11:881
79. Lifshitz IM, Grosberg AYu, Khokhlov AR (1978) Rev Mod Phys 50:683
80. Kuchanov SI (1992) Adv Polym Sci 103:1
81. Willert M, Landfester K (2002) Macromol Chem Phys 203:825
82. Moad G, Solomon DH (1995) The chemistry of free radical polymerization. Pergamon Press, London
83. Aris R (1965) Introduction to the analysis of chemical reactors. Prentice Hall, Englewood Cliffs, New Jersey
84. Kuchanov SI, Khokhlov AR (2005) Macromolecules 38:2937
85. Cox DR (1961) Renewal theory. Wiley, New York
86. Dotson NA, Galvan R, Laurence RL, Tirrell M (1996) Polymerization process modeling. Wiley, Cambridge, UK
87. Feller W (1966) An introduction to probability theory and its applications. Wiley, New York
88. Greenley RZJ (1980) Macromol Sci A 14:427,445
89. Korn GA, Korn TM (1961) Mathematical handbook for scientists and engineers. McGraw-Hill Book Company, New York
90. Bagdasaryan KhS (1966) Theory of radical polymerization. Nauka, Moscow
91. Kuchanov SI (1992) In: Allen G (ed) Comprehensive Polymer Science, First suppl, chap 2. Pergamon Press, New York, p 23
92. Kuchanov SI, Panyukov SV (1996) In: Allen G (ed) Comprehensive Polymer Science, second suppl, chap 13. Pergamon Press, New York, p 441
93. Stockmayer WH (1945) J Chem Phys 13:199
94. Kuchanov SI (1992) Adv Polym Sci 103:1
95. Kuchanov SI (1996) In: Kuchanov SI (ed) Mathematical methods in contemporary chemistry, chap 5. Gordon & Breach, Amsterdam, p 267
96. Odian G (1991) Principles of polymerization. Wiley, New York

Adv Polym Sci (2006) 196: 189–210
DOI 10.1007/12_055

Published online: 10 November 2005

After-Action of the Ideas of I.M. Lifshitz in Polymer and Biopolymer Physics

Alexander Yu. Grosberg[1,2] (✉) · Alexei R. Khokhlov[3,4,5]

[1]Institute of Biochemical Physics, Russian Academy of Sciences, 117977 Moscow, Russia
grosberg@physics.umn.edu

[2]School of Physics, University of Minnesota, 116 Church Street SE, Minneapolis, MN 55455, USA
grosberg@physics.umn.edu

[3]Physics Department, Moscow State University, 119899 Moscow, Russia

[4]Institute of Organoelement Compounds, Russian Academy of Sciences, 117823 Moscow, Russia

[5]Department of Polymer Science, University of Ulm, 89069 Ulm, Germany

Abstract We review the development of ideas in polymer physics initially formulated by I.M. Lifshitz. We start with general issues in polymer statistics, such as the analogy with quantum mechanics, and then continue with globules and a variety of coil-globule transitions, ranging from toroidal DNA particles, superabsorbing gels, to orientational order in liquid-crystalline polymers. We pay major attention to the more recent developments in the theory of heteropolymers with a quenched sequence of chemically distinct monomers and to the works on sequence design applied both in protein physics and in the attempts to mimic properties of proteins in synthetic copolymers.

Keywords Biopolymers · Copolymer sequences · Globules · Polymers · Sequence design

1
Introduction

The aim of this paper is to trace the scientific impact the ideas of I.M. Lifshitz had on polymer and biopolymer physics. We are not going to discuss the vast and, in our opinion, beneficial influence Iliya Mikhailovich had on the style, atmosphere, level, and productivity of research efforts of a great number of people. This multitude includes his own students as well as the studies of his students, his collaborators, people who consulted with him or simply attended his lectures and conference talks. We shall limit ourselves with the direct impact of his specific works on subsequent scientific literature.

Iliya Mikhailovich came to polymer science in the mid 1960s. This was a new field for him, though at that time he already had a number of outstanding achievements in many other branches of theoretical physics including, first of all, the physics of metals, crystals, and disordered systems. His interest in polymers was mainly due to the initial rapid advances of the emerging new science of molecular biology. In the 1960s, I.M. Lifshitz was among the most active participants of the so-called Schools on Molecular Biology. The latter were scientific meetings where many prominent physicists together with a large group of (then) young enthusiasts were learning this new science from each other. It was then that Iliya Mikhailovich came to the conclusion that biopolymers are in fact the level of biological structural hierarchy that can give rise to real biological physics, i.e., in particular, theoretical physics of biological systems, of animate matter. This conclusion was far from trivial for that time; moreover, it was rather a revolutionary one. The underlying idea was that a biopolymer molecule could be strictly described in terms of physics (even if this requires a lot of effort). At the same time, such a molecule is an information carrier and therefore is greatly endowed with that which can be somewhat vaguely described as "biological specifics".

This initial presumption called for the next step, namely, to concentrate attention on the fact that the binding of monomeric units into a polymeric chain is not governed by thermodynamic equilibrium conditions. On the contrary, the sequence of monomeric units in the chain embodies the memory of the (non-equilibrium) conditions of the synthesis. In particular, biological polymers are made in vivo through matrix or ribosomal synthesis that controls the primary sequences. Therefore, the importance of heteropolymers containing units of different types became clear right from the beginning. Nonetheless, the first paper on polymers I.M. Lifshitz published [1] was limited to homopolymers for simplicity's sake. However, he specially stressed the essentially non-equilibrium character of the linear structure even in homopolymers and the basic and inseparable connection existing therefore between polymers and his other favorite class of physical objects, namely, disordered systems.

Concentrating on the connection between polymers and disordered systems was a specific feature of the approach I.M. Lifshitz used in [1] and his other works. The other feature of paper [1], namely, clarifying the essential role of volume interactions between not immediately connected parts of the chain, places this work in line with the works of S.F. Edwards and P.-G. de Gennes also published during that period. These three researchers were those that made a breakthrough in polymer physics in the late 1960s by introducing modern techniques into this science and uncovering a wealth of new productive physical analogies. Their new ideas made polymers a respected topic among (most) physicists. Further development of these ideas led to the creation of a large new branch of physics called *soft condensed matter* physics.

2
Coil-Globule Phase Transition and Related General Issues of Polymer Theory

Technically, the central point of paper [1] was establishing a mathematical analogy between the statistical mechanics of a polymer chain and the quantum mechanics of a particle in an external potential field. S.F. Edwards [2] first noted this analogy shortly before I.M. Lifshitz did. Edwards has shown that the statistical sum over all the possible shapes of the polymer's spatial contour can be interpreted similarly to Feynman path integrals. I.M. Lifshitz rediscovered this analogy independently, and he paid greater attention to the eigenvalue equation, which is similar to Schroedinger's equation. The idea was that the choice of state for a polymer chain between a freely fluctuating coil or a densely packed globule depends on the presence of a discrete energy level. E. Cassaca [3] developed a similar technique in the context of polymer chain adsorbtion.

The very possibility of relatively dense states of polymer chains had been noted in the literature [4, 5] before the contribution by I.M. Lifshitz based on the following simple observation. In the case of a negative second virial coefficient (i.e., when monomers attract each other in pair collisions) Flory's theory predicts the coil size (characterized, e.g., by the chain's radius of inertia) somewhat less than the Gaussian one. However, I.M. Lifshitz raised this discussion to a much higher level in his paper [1]. He was the first to suggest the possibility of phase transitions within a single polymeric molecule, and the states of fluctuating coil and densely packed globule were first viewed as two distinct phases. These were the ideas that had the most clear and multiple implications, which we discuss below.

Since Flory's works [6] the following assumption became traditional in polymer physics. The sign change in the second virial coefficient occurs at such a temperature (or such a state of the solvent) that all the other virial coefficient's are positive: the monomer pairs can be attracted to each other, whereas repulsion is dominating in large groups of monomers. It so happened that I.M. Lifshitz considered the opposite case, namely, segment-segment attraction, in his paper [1]. This resulted in a seeming contradiction between the conclusions of I.M. Lifshitz and those of his predecessors; the issue was after all completely clarified in papers [7–9] published by I.M. Lifshitz and collaborators. The non-linear self-consistent equations first introduced by Lifshitz were later solved numerically for globules with a more traditional pair attraction [10] so that many aspects of the theory could be compared quantitatively with the experimental results.

P.-G. de Gennes later also considered the multisegment attraction regime. He suggested the so-called p-cluster model [11] in order to explain certain anomalies in behavior observed in many polymer species such as polyethyleneoxide (PEO); see also [12]. The scenario of coil-globule transition with dominating multisegment interaction first considered by I.M. Lifshitz has been recently studied in [13]. The authors used a computer simulation of chains in a cubic spatial lattice to show that collapse of the polymer can be due to crystallization within the random coil.

Concentrating on the coil-globule transition caused by multisegment interaction (i.e. "strong" first order transition) stimulated I.M. Lifshitz to take a vivid interest in the structure of the globule surface, where the local monomer density has an almost step-like change. In particular, the possibility of preserving mechanical equilibrium at the globule surface due to the non-local force transfer via the stretched polymer chain was suggested in [14]. It was shown in [15] (see also the review [8]) that no non-local force transfer is possible in a homopolymer globule. Similarly, no (macroscopic) electric field can exist in the bulk of metals, because it would otherwise have caused an electric current. Likewise, no stretched chain segments can exist (at a macroscopic scale), because such stretching would have caused reptation. Stretched chains, the entropy of their stretching, and non-local force transfer play an important role in systems with suppressed reptation. This can be the case, in particular, in heteropolymers, for example in the block-copolymer structures in a strong stratification regime, as A.N. Semenov first showed [16].

I.Ya. Erukhimovich evidences that I.M. Lifshitz formulated the problem of kinetics of the coil-globule transition immediately upon completion of his paper [1]. However, many years had passed before the results were obtained, and even now the question cannot be considered completely clarified. The qualitative idea I.M. Lifshitz suggested was that the transition should occur over many parts of the chain simultaneously [13]. O.B. Ptitsyn introduced a similar hypothesis [17] concerning the dynamics of protein globule collapse at approximately the same time. His hypothesis was as follows: fist secondary structure elements (e.g., α-spirals) are formed all over the chain and then all these elements fold together as already rigid blocks. G.-P. de Gennes considered this scenario for a homopolymer in his paper [18] that had a great impact on the subsequent works. Other authors also addressed this problem over many years. The most recent and advanced work in this direction was published in [19]. The experiments have shown that the hierarchical scenario described above is not often realized for proteins; moreover, such cases are rather rare (see, e.g., book [20] for details). This prompts the interest towards possible alternative scenarios for homopolymers. For example, recent computer simulation [21] has shown the possibility of nucleus formation during collapse of a hydrophobic homopolymer.

I.M. Lifshitz formulated the problem of the role of knots in the collapse of polymeric chains (in general) and in the self-organizing structures of protein globules (in particular) in his paper [14] published in 1973. He was the first to note that the issue of quasiknots (i.e., knots in a linear open chain) can be formulated more definitely in the case of compact polymer conformations, of globules, than in the case of open coils. Many authors have considered this idea since then (see, e.g. a recent paper [22]), and it has gradually transformed to the following understanding. The knots in a swollen coil have a statistical trend towards localization (i.e., they almost tighten in a certain part of the chain), whereas the knots are delocalized in a globule [23]. The issue of the number of knots in a globular chain was examined via computer simulation in papers [24, 25]. The influence of knots (or their absence thereof) on the dynamics of a coil-globule transition was studied in [26, 27]. The role of knots in the dynamics of reverse swelling of globules was considered in papers [28, 29]. The role of knots in conformational transitions in protein molecules was also studied extensively, but the issue still remains open. A recent review on this subject can be found in [30].

Note that the paper [1] by I.M. Lifshitz forestalled the corresponding experiments by at least 10 or rather 20 years. Experimental observation of such transitions is even now far from a routine procedure. Here we shall limit the discussion to mentioning several studies, which are in our opinion the most important achievements in this field [31–34]. We shall also refer to very informative reports on computer simulation of the coil-globule transition, namely, recent paper [35], and a very good reference list therein.

3
Other Phase Transitions of the Coil-Globule Type

As we mentioned earlier, I.M. Lifshitz was the first to realize that the coil-globule transition is not just a decrease of the chain size, but a phase transition to a condensed phase. Considering the multiplicity of known and possible condensed states, such a broad view on the coil-globule transition opened a perspective for unified understanding of a great number of physical phenomena in a variety of polymer systems. This gave rise to the concept of "coil-globule-type" transitions. Below we discuss several examples of such transitions.

The concept of the coil-globule transition plays an important role in understanding the compact states of double helix DNA both in vivo (in viruses, prokaryotic cells, and chromosomes) and in vitro. The latter implies man-made systems where DNA collapses in the presence of various condensing agents such as multivalent cations, organic solvents, water-salt solutions of neutral polymers, surfactants, etc. [36]. The DNA collapse is difficult to understand, because it is not easy to imagine a strong enough mechanism of attraction between DNA segments to overcome their repulsion due to the very significant negative charge. The nature of attraction between DNA systems in the presence of multicharged cations (which is practically the most important case) was clarified in [37]. Adding a neutral polymer has a more limited action and can cause DNA collapse only in the presence of a sufficient concentration of salt that screens the charge of DNA itself (via Debye screening) [38]. Fluorescent microscopy has been used recently for studying DNA condensation, since this technique allows the tracing of a single DNA molecule. With this method significant new experimental insights were obtained. For example, it was found that a transient region (for precipitating agent concentration) exists in the process of DNA folding. A strongly fluctuating coil and dense globule can co-exist in this case [34, 39, 40]. The state diagram in this region is bimodal, and therefore the coil-globule transition itself is a first order phase transition.

The "globules" of such polymers as double helix DNA have a remarkable shape, namely the toroid one, that is they have a hole in the center. (Since the word "globule" originates from the Latin *globus*, i.e. sphere, calling such a DNA conformation globule without quotation marks is somewhat funny, though this is customary in scientific literature.) The possibility of DNA collapse to the toroid globules was theoretically predicted already during the lifetime of I.M. Lifshitz [41]. Several studies performed later were dedicated to finding the equilibrium size of toroid DNA globules [42]. However, this issue is still open, especially for toroids formed by several double helical DNAs. There is still no understanding of the reasons for toroid growth termination, and whether the observed toroid size is in fact the equilibrium one, or the growth is arrested kinetically.

The DNA conformation in vivo is also globular (meaning collapsed). In particular a lot of effort was spent on studying the structure of a DNA globule packed in the virus (bacteriophage) head. The relative simplicity of the problem makes such studies most promising. See [43] on the first steps in this direction, [44] or reviews [45, 46] on the recent progress, and [47] on DNA reptation from the virus when it attaches the cell. More risky attempts were also made to describe phenomenologically the most complex DNA in chromosomes of eukaryotic cells. These attempts led to formulation of the hypothesis of fractal hierarchical structure of chromosomes [48, 49]. Though the hierarchical character of the chromosome structure is qualitatively beyond doubt, the more detailed questions concerning this structure (as to the fractal dimension) are yet to be answered. The role of topology and DNA charge is not clear, either.

Yet another fascinating story developed as the attempts were made to study flexible polymers containing charged units dissociated in water. This class of polymers is very important from the practical point of view. Such molecules are called weakly charged polyelectrolytes. The dissociation degree of such molecules is high enough to ensure good solubility in water, but the number of charges is small enough, so that hydrophobic and Van der Waals interactions could compete with the Coulomb repulsion. First, it was found [50] that in a poor (for neutral units) solvent the molecules of weakly charged polyelectrolytes form an oblong "globule". This led to an assumption that the oblong globule has a "cucumber-like" form. Later the picture was found to be still more interesting and similar to the instability of a charged liquid droplet discovered by Raleigh. The stretched globule is in fact more like a string of onions or beads—small sub-globules tied together with stretched polymer segments [51]. Such "necklaces" have quite unusual properties. For instance, phase transitions are possible in such molecules connected with redistribution of matter between the sub-globules and the interconnecting segments of the chain [52]. Another class of polymers with similar properties are polyampholytes, which contain sequences of both positive and negative units. It was also found that a typical conformation for such molecules (i.e. that realized for a non-correlated sequence of positive and negative units with net uncompensated charge of the order of $N^{1/2}$) is not just a stretched globule [53], but a necklace containing many sub-globules [54].

The ideas I.M. Lifshitz suggested in his works were also used in developing a theory of another related phenomenon, namely, microphase separation in polyelectrolyte systems [55].

Yet another remarkable coil-globule transition has been found recently [56, 57]. The systems that have such notable properties can in fact often be met among polymers: each monomeric unit contains a hydrophobic group belonging to the chain itself, and a pendant hydrophilic group. Thus, the monomeric units themselves are amphiphilic. Note that most aminoacids have such properties. The number of possible conformations in such macromolecules is much greater than predicted based on the current models of homopolymeric chains. The globule of such a macromolecule (containing hydrophobic-hydrophilic units) up to a pretty large molecular mass has a cylindrical form, and its radius of inertia depends almost linearly on the polymerization degree N (just like for a completely stretched chain) over a wide range of N values. The transition between a coil and the cylindrical globule in amphiphilic macromolecules can also pass the "necklace" stage. Similar to many water-soluble globular proteins, copolymers containing amphiphilic units can form stable globules in the solution, which do not aggregate [56, 57].

Observing experimentally the collapse of separate polymeric chains is difficult, because collapse is usually accompanied by the tendency of chains to stick together and precipitate. On the other hand, collapse and swelling of a macroscopic polymeric chain, i.e. gel, can be quite easily observed with the naked eye. This phenomenon was discovered rather long ago, and is traditionally interpreted as a macroscopic manifestation of coil-globule transition. A vast amount of research literature is devoted to this effect (see, e.g., [58]) in view of the variety and importance of its technological applications (beginning with disposable baby diapers and ending with water treatment technologies). Polyelectrolyte gels can undergo the most drastic changes during the collapse process. The linear size of such macromolecules can change by a factor of tens or even hundreds. The electrostatic interactions proper are remarkably small in such gels, and the osmotic pressure of counterions is the main factor of the transition. The counterions are formed via dissociation of charged groups of the polymeric network and are "trapped" within the network sample due to the conditions of net electroneutrality of the sample [59].

The transitions of the coil-globule type were considered not only in the usual space, but also in the space of monomeric units' orientation, where such transition is equivalent to the nematic liquid crystal ordering [60, 61]. Such an approach using the formalism developed by I.M. Lifshitz has led to the creation of the theory of liquid crystal ordering in the solutions of semi-flexible macromolecules [62, 63].

4 Heteropolymers

Already at the very beginning of his studies in this field [1] I.M. Lifshitz insisted that the most important stage of the work to follow should be developing a theory of heteropolymers with a frozen (or quenched) disordered sequence of chemically different units. Many years passed before an essential progress was made in this direction despite the motivation originating from protein physics, that is the very problem of protein globules that I.M. Lifshitz had in mind while starting to study polymers.

The first essential step was made in paper [64], where the phenomenological considerations yielded the insight into the connection of the heteropolymer globule problem with the so-called random energy model (REM) developed previously in the context of spin glasses [65]. The next important achievement was the work [66] that takes its origin in the ideas developed by I.M. Lifshitz. In contrast to the authors of [64], the authors of [66] developed a microscopic theory of globular state of heteropolymers with a quenched inhomogeneous sequence of units.

The main idea of [66] was to consider a model in a sense even more random than just a random sequence of a limited (small) number of monomer types. Instead of considering just N independent random monomers, the authors assumed there are $N(N-1)/2 \sim N^2 \gg N$ independent arbitrary energies of interaction between the monomers (which means in fact that the interaction energy is different for each pair of N different monomers constituting the macromolecule). Note also that the principle of self-averaging of free energy was an important component of the study [66] and all the subsequent papers on the subject. I.M. Lifshitz paid much attention to this principle in the context of physics of disordered systems [67].

Many other models were studied in the course of the subsequent research, which has led to an understanding that REM is a correct approximation for low-energy states of heteropolymer globules. The grounds for the validity of this model lie in the basic properties of the globule as such, the very properties that I.M. Lifshitz discovered.

The connectivity of the chain (which I.M. Lifshitz called "linear memory") and excluded volume taken together impose such rigid limits on the conformation of globular polymers that the number of allowed local transformations of the chain is very small. Therefore, different low-energy states are separated by a global (and not by a local) transformation, that is there is a strong structural distinction between them. This, in turn, makes them statistically almost independent, which implies the validity of REM (see the review [68] for details). Thus, in addition to the two homopolymer phases I.M. Lifshitz discovered (namely, coil and globule), the random heteropolymer has yet a third phase—the frozen globule dominated by a single conformation (leaving microscopic fluctuations aside). First, it was supposed that the frozen phase is a reasonable model for a "correctly" collapsed protein, but this hypothesis soon proved to be wrong, because the frozen phase of a random heteropolymer has no stability, either against mutations, or against small variations of the properties of the polymer's environment (e.g., ionic strength, etc.).

5 Sequence Design

The invalidity of the random heteropolymer model for proteins can also be understood from a more general point of view, since we know that protein sequences result from long evolution and are therefore, strictly speaking, not random. This consideration was the origin of a very productive approach that considers heteropolymer physics in connection with their evolution. This idea was not mentioned in any published work by I.M. Lifshitz. However, the authors of the present paper can witness that I.M. Lifshitz had "evolutionary" ideas on this subject and often discussed them.

An essential progress in this direction was achieved [69–71] with the following reformulation of the problem. Instead of trying to solve a very complex problem of real biological evolution, an attempt was made to find a way of preparing sequences ensuring certain properties of the synthesized heteropolymer. This problem is called polymer sequence design. Generally speaking, sequence design can be considered as an approach to creating copolymers having required structural and/or functional properties. Conformation-dependent design is one of the most promising directions of copolymer synthesis. Its main idea consists in finding a substitute for the typically used technique of sequential monomer addition to a one-dimensional sequence (which reminds one of the well-known Eddington's parable of a monkey typing at the typewriter). Instead, it is suggested to use a polymer already in a globular conformation and to form the sequence directly in the globule. As I.M. Lifshitz explained, each monomer in a globular macromolecule interacts almost exclusively with other units except his immediate neighbors in the chain. The conditions of synthesis or chemical modification of the copolymer, including the particular globular structure, are (in some way or other) coded (recorded, imprinted) in the primary sequence of this copolymer. Thus, this sequence can be considered as a text containing characters from a finite alphabet, i.e. monomer units of different types. In other words, the resulting sequence inherits some properties of the template ("parent") conformation, which the copolymer had in the course of synthesis (modification).

The issue of information content in sequences prepared with a conformation-dependent design technique is of great interest. Intuition makes it clear that the information content in the sequences of synthesized polymers is much lower than that in DNA that carries the complete genetic information concerning the organism or in a globular protein tuned to perform a single strictly defined function. Nonetheless, conformation-dependent design allows the writing of much more information in the polymer sequences than the more traditional techniques of chemical synthesis. This statement is true even in the case when only two types of units are used in conformation-dependent design, for example hydrophilic and hydrophobic molecules (two-symbol alphabet) [69–72]. Note besides that all the macromolecules of a given biological polymer are identical due to the perfection of the biosynthetic apparatus of the cell. Therefore, each such molecule carries complete information concerning its function, whereas the "conformation memory" of synthetic copolymers produced with a conformation-dependent design is a property averaged over a molecular ensemble. The molecules in the ensemble, though prepared under identical conditions, are, however, more or less different, and their similarity is only statistical.

This branch of polymer physics is closely connected to another of I.M. Lifshitz favorite directions in physics, namely, the theory of disordered systems [73]. The situation when different samples have only statistical similarity is typical for the physics of chaos, and many concepts I.M. Lifshitz developed are also quite naturally applied in the physics of disordered polymers. The idea of self-averaging in general and self-averaging of free energy [74], in particular, are examples of such concepts.

One of the specific features of biology is the possibility of using a living cell in order to produce a macroscopic number of perfectly identical disordered systems (and not just statistically similar systems). This feature of biology opens new perspectives for physics (and for physicists), but, as usual, also presents new challenges.

6
Sequence Design and Protein Physics

If the template conformation is a globular one, then the conformation-dependent design procedure yields sequences with strongly non-local correlations along the chain (in contrast to the "monkey-at-typewriter" technique). This is due to the properties of the globule. Note that as soon as this fact was perceived, the corresponding correlations were found in real proteins [76], as well as in artificially designed sequences [75, 77]. Correlation statistics in sequences created with one of the design techniques satisfy a remarkable power law, that is these sequences have Levy-flight-type long-range correlations [78]. Obtaining correlations of a certain type in the chain is not an end in itself, of course. The interesting physical properties of heteropolymers with specially designed sequences are more important. In many cases these properties are similar to those of proteins, therefore the corresponding synthetic macromolecules are often called protein-like polymers.

The most remarkable property of protein molecules is the ability to spontaneously collapse into a definite spatial structure, which is always the same for a given molecule, and which is encoded in the sequence of the units constituting the protein. Therefore, the following question was the first within the framework of the conformation-dependent design program: is it possible to design a sequence such that the synthesized macromolecule has this very important property of proteins? This question can be subdivided into the two following problems. First, can a well-pronounced global minimum of energy be ensured for a given conformation of the polymer molecule? Second, is it possible to ensure that, apart from having this equilibrium state, the molecule could reach it kinetically from an arbitrary other state? The answers to both these questions are positive, and the corresponding phenomenon was realized in computer experiments for lattice models of polymers even for relatively long chains containing up to 100 units [79]. Despite the relative novelty of the topic, an extensive literature exists on this issue, which is impossible to review within the present paper. Therefore, we recommend our reader to study the detailed bibliography given in the reviews [68, 80–85]; and also the book [20].

Though several design schemes were suggested to produce uniquely collapsing copolymers, the main idea of all the approaches is generally the same. This idea consists of finding a minimum of the chain energy $E(C, S)$ (which depends on conformation C and sequence S), with respect to sequence S for a given conformation C. This coincides exactly with the above-discussed scheme of conformation-dependent design, because it corresponds to the process where chemical formation of the sequence occurs based on a frozen or slowly changing conformation. An important observation here is that the relaxation dynamics of the system in the sequence space for a frozen conformation (in contrast to the conformational dynamics of a heteropolymer with a frozen sequence) is in no danger of frustration. Therefore, this relaxation is fast and reliable and cannot get stuck at any local minimum. Theory shows (see the reviews mentioned above) that the frozen phase of polymers with designed sequences (in contrast to that of the random ones) is not a random, but a certain given conformation. The reason for this is the relative rarity of local transformations of the globule and in statistical independence of low-energy states. This implies that the sequence can be chosen so as to reduce the energy of the single state (by a thermodynamically significant value), but at the same time to leave intact the statistics of all the other (low-energy) states.

Polymers designed with this technique have a number of important aspects in common with proteins. First of all, the transition from a liquid-like globule into a frozen state occurs as a first order phase transition. Further, the frozen state itself has an essential stability margin, which is determined by the design parameters. As in real proteins, neither a large variation of temperature or other environmental conditions, nor a mutational substitution of several monomers leads to any change in basic state conformation. In this respect the ability of sequence design to capture certain essential characteristics of proteins seems quite plausible.

Certain difficulties remain, however, with this approach. First, such an important feature as a secondary structure did not find its place in this theory. Second, the techniques of sequence design ensuring exact reproduction of the given conformation are well developed only for lattice models of polymers. The existing techniques for continuum models are complex, intricate, and inefficient. Yet another aspect of the problem is the necessity of reaching in some cases beyond the mean field approximation. The first steps in this direction were made in paper [84], where an analog of the Ginzburg number for the theory of heteropolymers was established.

In the meantime, while theoreticians are still working on developing and perfecting their approaches, an essential progress was reached in more empirical techniques. Empirical methods of computer-aided sequence design are in principle similar to those applied for the lattice models [68, 80–83], that is they are directed at building a sequence with a minimum possible energy for a given conformation. We are not giving here references to the vast amount of works published on the topic, but cannot help mentioning the following outstanding achievement. The authors of a recent paper [86] achieved building a sequence of 93 aminoacids that has minimum energy for a given spatial structure called *Top7*. Then, after adding 13 more aminoacids to the ends of the molecule (which was necessary for expression of the latter) and following purification, the corresponding "gene" was really synthesized. This synthesis yielded artificial "protein" molecules, which collapse uniquely, correctly, and reversibly to the given globular form. Moreover, these globules were later crystallized and subjected to X-ray structural analysis, which confirmed that the precision (root mean square deviation) of reconstructing the "planned" globular structure was better than 1.2 Angstrom. The most remarkable in this study is that the *Top7* spatial structure is a completely novel, fully artificial one that is never found among real proteins in nature.

Attempts were also made to design non-protein polymers able to collapse to a globular conformation with properties complying (at least in part) with the given requirements e.g., having an active center [87]. The review of the research performed in this direction can be found in [88–92].

The above-mentioned issue of information content in the heteropolymer sequence has interesting implications in the context of protein physics. If the molecule can collapse to a certain spatial structure, then saying that the information about this conformation is contained (recorded) in the sequence would be reasonable. The next question is whether the sequence can remember two certain spatial structures instead of just one? In general, how many spatial structures can be recorded in a single polymer sequence? The answer to this question was recently found in paper [94], and it is based on a deep analogy between heteropolymer sequences and neuron networks [93]. The rather unexpected conclusion was made that the maximum number of conformations recorded in the sequence does not depend on its length N (for large N), but is approximately $\ln Q$, where Q is the number of chemically different types of units (symbols in the sequence alphabet). This result explains directly why proteins or uniquely collapsing protein-like copolymers cannot be built of two types of units (as any theoretician would have preferred): $\ln 2 = 0.7 < 1$. On the other hand, increasing the "richness of the monomer alphabet" Q ensures better stability (both mutational and thermodynamic) of the ground state [95–97]. At least 5–6 monomer types are necessary to ensure a minimal reasonable reproducibility of collapse [98–100]. These dependences have interesting implications. The spontaneous collapse of proteins to a wrong conformation can cause diseases of animals and humans (e.g., bovine spongiform encephalopathy (BSE), the so called "mad cow disease"). Such diseases can be an evolutionary price of the trade-off between the tendency of making proteins more stable (which requires higher Q) and the attempts to prevent errors in protein folding (that requires lower Q).

Another (quite different) implication of the issue of information capacity of protein-like copolymer is the possibility of recording a small region in conformation space near the native state instead of recording several essentially different conformations in the polymer sequence. Moreover, this can be done so that the transitions between the conformations within the above-mentioned region occur fast, without barriers, and possibly in an ordered way, as is done in functioning natural proteins. Models of this type were studied in paper [101].

7 Sequence Design for Synthetic Polymers

Other requirements can be imposed on sequence design, not necessarily connected with the unique folding of a polymer chain into a globule having a certain structure and minimal possible energy. The principles of conformation-dependent design have led to the development of new synthesis strategies aimed at production of functional copolymers (see [102, 103] and recent reviews [104–106]). These strategies can be divided into two groups: (1) polymer-analogous transformation (chemical modification) of a part of the units in macromolecules; and (2) copolymerization of monomers with different properties performed under special conditions.

In order to perform polymer analogous transformation the conditions should be ensured in the reaction system that make the original homopolymeric chain assume a certain initial conformation (globular or adsorbed one). A part of the units in the macromolecule is screened from the solution in this case, while a reagent introduced into the system chemically modifies the other part of the units. Thus, the initial (parent) conformation of the chain is used as a certain template for the copolymer to be obtained. The following particular instance of copolymer unit design was used in a large series of studies [57, 71, 107–114] initiated with the papers [102, 103]. The design of copolymer sequences in these studies was realized in such a way as to reproduce one of the fundamental properties of many proteins, namely, solubility in water and aggregative stability of their globular conformation. The corresponding copolymers forming globules with a hydrophobic core and hydrophilic shell are in a certain way protein-like. The technique for synthesizing such molecules has a biomimetic nature, which implies the following three stages. First, the principal properties of the initial biopolymer sequences are determined, second, understanding is reached as to how these sequences impact the functional properties of the polymers, and then the acquired knowledge is used for synthesizing new copolymers. If a partially adsorbed homopolymeric chain is taken as the "parent" conformation, then polymer-analogous modification lead to the creation of specific copolymers preferring adsorption onto a plane or colloid particles of a certain size [114–118]. This opens a way to creating a so-called molecular dispenser [118]. Theoretical studies of copolymers produced via chemical modification of homopolymeric globules and adsorbed chains have shown that the resulting sequences are characterized by remarkable long-range correlations (which are unusual for typical random copolymers) and have Levy-flight statistics. It was found also that thus designed protein-like copolymers have an enhanced ability to self-organization. Moreover, this effect is observed not only in a dilute solution during the formation of microsegregated globular structures of the "core-shell" type, but also in melts [119, 120]. Unusual self-organization forms were found in protein-like copolymer systems containing electrically charged hydrophilic groups [111–113].

Conformation-dependent copolymerization is very similar to the well-known matrix polymerization [105, 106]. This sequence design technique can be considered as editing the texts typed by Eddington's monkey during the typing process itself. Spatially inhomogeneous concentration fields of the monomers being polymerized play here the role of "editors". Due to the presence of concentration gradients in the reaction mix, the sequence and equilibrium conformation of the growing macroradical become interdependent and determine the order of monomer addition. One of the first attempts to realize this approach was based on copolymerization of monomers with different solubility accompanied by globule formation [121, 122]. Techniques were developed for conformation-dependent template copolymerization near a homogeneous surface that selectively adsorbs monomers of a certain type from the reaction mix [115–117], and also near a surface with a given distribution of adsorbtion centers [123]. Thus, copolymers with rather exotic primary sequences can be designed (e.g., gradient ones and others). Copolymers able to discern microscopic patterns on the surface can also be produced with this technique [123, 124].

Corresponding approaches were developed in all the research methods: theoretical, computer simulation, and, moreover, experimental. Thus, copolymers were synthesized in vitro, which form non-aggregating structures of the type hydrophobic core–hydrophilic shell. The structure of such copolymers is similar in this respect to that of protein macromolecules [125–127].

All the examples described above clearly show that the fundamental theoretical ideas of I.M. Lifshitz find more and more practical realizations.

8 Melting of Heteropolymer DNA

One of the papers of I.M. Lifshitz (namely, [128]) stands somewhat apart from his other works. In this paper the problem of the melting of a double helix DNA molecule with a given quenched sequence of aminoacid units was solved in the most general form. At that time, when the paper was published, researchers hoped to find a way of "reading" the DNA sequences using temperature dependences of their melting. However, this hope was not realized, and nowadays DNA sequences are determined with other techniques. Nonetheless, the work [128] outlived its motivation. The main idea of the work of mapping the sequence on the random walk trajectory is widely applied (see, e.g., [129]). A recent study [130] of "unzipping" of the DNA double helix under a constant external force applied to the ends of the chain (via optical tweezers) is based (both in ideological and methodical sense) on the paper [128].

9 General Questions of Biophysics

The paper of I.M. Lifshitz [1] caused an immediate and enthusiastic response from the trendsetters in biophysics in the USSR [131, 132]. This enthusiasm did not fade with the passing of years. Even in the last book of L.A. Blumenfeld recently published [133] the theory developed by I.M. Lifshitz was ranked among the three or four main achievements of biological physics in the twentieth century. The basics of the theory of coil-globule transitions are treated in the well-known course of biophysics [134] and thus acknowledged as fundamental for this science. Here, it is necessary to point out that similar level courses of biophysics published in the West do not mention the theory of I.M. Lifshitz. This is mainly due to the Western scientific tradition, where attention is mostly paid to the specific subject of the paper rather than to the general approach formulated therein. In the case of paper [1] the specific subject was the homopolymer; i.e., strictly speaking, it was a non-biological object.

The specific fundamental idea formulated in paper [1], which found most applications in biophysics, is the concept of incomplete equilibrium, of kinetically frozen degrees of freedom. This idea is at the base of modern concepts of enzyme functioning [135–137], and in many cases it is clearly formulated as one originating from the paper [1] of I.M. Lifshitz.

In fact, the great response to the paper [1] was caused not by the paper alone and the ideas immediately formulated therein, but rather with the many presentations I.M. Lifshitz made on the subject of the paper at various seminars. His subsequent reports, comments, and remarks at seminars not directly connected with his paper [1] also contributed to the impact of his work. For example, the discussion of the well-known paper of M. Eigen at the seminar of M.V. Volkenstein, the discussion of the report made by A.A. Neufach at the seminar of P.L. Kapitza, and many others can be mentioned here.

As we have already mentioned, paper [1] resulted from the particular interest I.M. Lifshitz had in biological physics. Iliya Mikhailovich himself has always considered this study and presented it as the first step towards "real theoretical physics of biological systems" (here a fragment of one of the public presentations of I.M. Lifshitz is quoted). I.M. Lifshitz always explained in his reports what a potential of biological generalization his paper [1] contains, and how the next steps in this direction can and should be done. We have tried several times to retell in our own words the ideas Iliya Mikhailovich formulated in his presentations (see, e.g., [139]). Here we believe it suffices to say that the statements of I.M. Lifshitz concerning the physical basis of biology, and especially those concerning evolution were profound and well reasoned, and, what is most important, very exciting and inspiring. The studies of sequence design described above, though dating many years after the departure of I.M. Lifshitz, were nonetheless the direct result of inspiration he radiated.

After the sequence design analysis an essential progress was made in understanding the evolution at the molecular level. This problem was always of great interest to I.M. Lifshitz. The term *designability of conformation* widely used in the English-speaking scientific community is a very important concept for this problem. Designability of conformation for a conformation C is defined as the number of sequences, which have the minimum energy in this conformation. This is in fact the analogue of entropy in the sequence space. This notion was first introduced by A.V. Finkelstein (see [20, 140]) and later developed (and given its present name) in paper [141]. Without going into detail, we can only remark that conformations with high designability play an important role in the evolution understood as a cooperative diffusion in the sequence space (see, e.g., [142] and the review [143]). Note also that various scenarios of sequence evolution were considered for the models of protein-like copolymers [144–147].

10 Conclusion

Looking back, we must state with a note of regret that the contribution by I.M. Lifshitz to the physics of polymers and biopolymers, however widely known and esteemed in the scientific community, is still known and esteemed insufficiently. Many of his ideas were not understood and had to be rediscovered, and his priority is not always mentioned, as it should have been. Especial regret is felt because the more general ideas that I.M. Lifshitz often presented and discussed in public, but never published in printed form, are practically unknown to the scientific community. Today, researchers are not used to presenting ideas as yet unpublished, and most people believe that the achievements of a scientist should be rightly judged by his printed works alone. This, however, would be most unjust when applied to I.M. Lifshitz.

Though we believe that the priority of I.M. Lifshitz should be mentioned more often and expressly, Science itself cares not for such priority, and only the history of ideas is of essence. However, we cannot but come to the conclusion that the ideas I.M. Lifshitz developed in polymer physics had, generally, good fortune. They were really innovative, proven as true, and had left a clear trace. We think, I.M. Lifshitz would have been content.

References

1. Lifshitz IM (1968) Some problems of the statistical theory of biopolymers. Zh Eksp Teor Fiz 55:2408; [(1969) J Exp Theor Phys 28:1280]. See also Lifshitz IM (1994) In: Selected scientific works. Electronic theory of metals. Polymers and biopolymers. Nauka publishers, Moscow, p 212
2. Edwards SF (1965) Proc Phys Soc 85:613
3. Cassassa EF (1967) J Polym Sci B5:773
4. Stockmayer WH (1960) Macromol Chem Phys 35:54
5. Ptitsyn OB, Eisner YuE (1965) Biofizika 10:3
6. Flory PG (1953) Principles of polymer chemistry. Cornell University Press, Ithaca, NY
7. Lifshitz IM, Grosberg AYu, Khokhlov AR (1976) Zh Eksp Teor Fiz 71:1634; [J Exp Theor Phys 44:855 (1976)]; See also Lifshitz IM (1994) In: Selected scientific works. Electronic theory of metals. Polymers and biopolymers. Nauka publishers, Moscow, p 261 (in Russian)
8. Lifshitz IM, Grosberg AYu, Khokhlov AR (1978) Rev Mod Phys 50:683; See also Lifshitz IM (1994) In: Selected scientific works. Electronic theory of metals. Polymers and biopolymers. Nauka publishers, Moscow, p 270 (in Russian)
9. Lifshitz IM, Grosberg AYu, Khokhlov AR (1979) Usp Fiz Nauk (Soviet Physics—Uspekhi) 127:353; See also Lifshitz IM (1994) In: Selected scientific works. Electronic theory of metals. Polymers and biopolymers. Nauka publishers, Moscow, p 299 (in Russian)
10. Grosberg AYu, Kuznetsov DV (1992) Macromolecules 25:1970; Macromolecules 25:1980; Macromolecules 25:1991; Macromolecules 25:1996
11. de Gennes PG (1991) CR Acad Sci Paris II 313:1117
12. Baulin VA, Halperin A (2003) Macromol Theory Simul 12:549
13. Rampf F, Paul W, Binder K (2005) Europhys Lett 70:628
14. Lifshitz IM, Grosberg AYu (1973) Zh Eksp Teor Fiz 65:2399; [(1974) J Exp Theor Phys 38:1198]; See also Lifshitz IM (1994) In: Selected scientific works. Electronic theory of metals. Polymers and biopolymers. Nauka publishers, Moscow, p 236 (in Russian)
15. Lifshitz IM, Grosberg AYu (1975) Dokl Phys 220:468; See also Lifshitz IM (1994) In: Selected scientific works. Electronic theory of metals. Polymers and biopolymers. Nauka publishers, Moscow, p 257 (in Russian)
16. Semenov AN (1985) Zh Eksp Teor Fiz 88:1242 [J Exp Theor Phys 61:733]
17. Ptitsyn OB (1973) Sequential mechanism of protein folding. Dokl Phys 210:1213 (in Russian)
18. de Gennes PG (1985) J Phys Lett 46:L639; See also Buguin A, Brochard-Wyart F, de Gennes PG (1996) CR Acad Sci Paris IIb 322:741
19. Abrams CF, Lee NK, Obukhov S (2002) Europhys Lett 59:391
20. Finkelstein AV, Ptitsyn OB (2004) Protein physics. A course of lectures. Academic Press, New York
21. ten Wolde PR, Chandler D (2002) Proc Natl Acad Sci USA 99:6539
22. Millett K, Dobay A, Stasiak A (2005) Macromolecules 38:601
23. Markone B, Orlandini E, Stella AL, Zonta F (2005) J Phys A: Math Gen 38:L15; Orlandini E, Stella AL, Vanderzande C (2004) J Stat Phys 115:681; Orlandini E, Stella AL, Vanderzande C (2003) Phys Rev E 68:031804
24. Mansfield ML (1994) Macromolecules 27:5924
25. Lua R, Borovinskiy A, Grosberg A (2004) Polymer 45:717
26. Grosberg AY, Nechaev SK, Shakhnovich EI (1998) J Phys (France) 49:2095

27. Grosberg AY, Kuznetsov DV (1993) Macromolecules 26:4249
28. Rabin Y, Grosberg AY, Tanaka T (1995) Europhys Lett 32:505
29. Lee NK, Abrams CF, Johner A, Obukhov S (2003) Phys Rev Lett 90:225504
30. Taylor WR, May ACW, Brown NP, Aszodi A (2001) Rep Prog Phys 64:517
31. Nishio I, Sun ST, Swislow G, Tanaka T (1979) Nature 281:208; Swislow G, Sun ST, Nishio I, Tanaka T (1980) Phys Rev Lett 44:796
32. Chu B, Ying Q (1996) Macromolecules 29:1824; Chu B, Ying Q, Grosberg AY, Macromolecules 28:180
33. Wu C, Zhou S (1996) Phys Rev Lett 77:3053; Nakata M, Nakagawa T (1997) Phys Rev E 56:3338; Wu C, Wang X (1998) Phys Rev Lett 80:40921998; Zhang G, Wu C (2001) Phys Rev Lett 86:822
34. Yoshikawa K, Takahashi M, Vasilevskaya VV, Khokhlov AR (1996) Phys Rev Lett 76:3029
35. Polson JM, Moore NE (2005) J Chem Phys 112:024905
36. Bloomfield VA (1996) Curr Opin Struct Biol 6:334
37. Rouzina I, Bloomfield VA (1996) J Phys Chem 100:9977
38. Grosberg AYu, Erukhimovich IYa, Shakhnovich EI (1982) Biopolymers 21:2413; Grosberg AY, Zhestkov AV (1986) J Biomol Struct Dyn 3:859
39. Yoshikawa K, Kidoaki S, Takahashi M, Vasilevskaya VV, Khokhlov AR (1996) Ber Bunsenges Phys Chem 100:876
40. Takahashi M, Yoshikawa K, Vasilevskaya VV, Khokhlov AR (1997) J Phys Chem 101:9396
41. Grosberg AY (1979) Biofizika 24:32
42. Vasilevskaya VV, Khokhlov AR, Kidoaki S, Yoshikawa K (1997) Biopolymers 41:51; Stukan MR, Ivanov VA, Grosberg AY, Paul W, Binder K (2003) J Chem Phys 118:3392
43. Gabashvili IS, Grosberg AY, Kuznetsov DV, Mrevlishvili GM (1991) Biofizika 36:780
44. Kindt J, Tzlil S, Ben-Shaul A, Gelbart WM (2001) Proc Natl Acad Sci USA 98:13671; Lambert O, Letellier L, Gelbart WM, Rigaud (2000) Proc Natl Acad Sci USA 97:7248
45. Gelbart W, Bruinsma R, Pincus P, Parsegian A (2000) Physics Today 53:38
46. van der Schoot P, Bruinsma R (2004) preprint q-bio.BM/0410032 (http://xxx.lanl.gov/ftp/q-bio/papers/0410/0410032.pdf)
47. Gabashvili IS, Grosberg AY (1992) J Biomol Struct Dyn 9:911
48. Grosberg AY, Rabin Y, Havlin S, Neer A (1993) Europhys Lett 23:373
49. Sikorav JL, Jannink G (1993) CR Acad Sci Paris 316:751
50. Khokhlov AR (1980) J Phys A: Math Gen 13:979
51. Dobrynin AV, Rubinshtein M, Obukhov SP (1996) Macromolecules 29:2974
52. Dobrynin AV, Rubinshtein M (1999) Macromolecules 32:915
53. Gutin AM, Shakhnovich EI (1994) Phys Rev E 50:R3322; Dobrynin AV, Rubinshtein M (1995) J de Phys II (France) 5:677
54. Kantor Y, Kardar M (1991) Europhys Lett 14:421; Kantor Y, Li H, Kardar M (1992) Phys Rev Lett 69:61; Kantor Y, Kardar M, Li H (1994) Phys Rev E 49:1383; Kantor Y, Kardar M (1994) Europhys Lett 27:643; Kantor Y, Kardar M (1994) Europhys Lett 28:169; Kantor Y, Kardar M (1995) Phys Rev E 52:835; Kantor Y, Kardar M, Ertas D (1998) Physica A 249:301
55. Erukhimovich IYa, Dobrynin AV (1992) Macromolecules 25:4411; Khokhlov AR, Nyrkova IR (1992) Macromolecules 25:1493
56. Vasilevskaya VV, Khalatur PG, Khokhlov AR (2003) Macromolecules 36:10103
57. Vasilevskaya VV, Klochkov AA, Lazutin AA, Khalatur PG, Khokhlov AR (2004) Macromolecules 37:5444
58. Dusek K (1993) Adv Polym Sci 109:110

59. Khokhlov AR, Starodubtsev SG, Vasilevskaya VV (1993) Adv Polym Sci 109:123
60. Khokhlov AR, Semenov AN (1981) Physica A 108:546
61. Khokhlov AR, Semenov AN (1982) Physica A 112:605
62. Khokhlov AR (1988) In: Plate NA (ed) Liquid crystal polymers. Khimia, Moscow (in Russian)
63. Semenov AN, Khokhlov AR (1988) Physics Uspekhi 156:427
64. Bryngelson JD, Wolynes PG (1987) Proc Nat Acad Sci USA 84:7524
65. Derrida B (1980) Phys Rev Lett 45:79
66. Shakhnovich EI, Gutin AM (1989) Biophys Chem, vol 34, p 187
67. Lifshitz IM, Gredeskul SA, Pastur LA (1988) Wiley, New York
68. Pande VS, Grosberg AY, Tanaka T (2000) Reviews of Modern Physics 72:259
69. Shakhnovich EI, Gutin AM (1993) Proc Nat Acad Sci USA 90:7195
70. Pande VS, Grosberg AY, Tanaka T (1994) Proc Nat Acad Sci USA 91:12972
71. Khokhlov AR, Khalatur PG (1999) Phys Rev Lett 82:3456
72. Lau KF, Dill KA (1989) Macromolecules 22:3986
73. Lifshitz IM, Gredeskul SA, Pastur LA (1984) Nauka, Moscow (in Russian)
74. Chuang J, Grosberg AY, Kardar M (2001) Phys Rev Lett 87:078104
75. Pande VS, Grosberg AY, Tanaka T (1994) J Chem Phys 101:8246
76. Pande VS, Grosberg AY, Tanaka T (1994) Proc Nat Acad Sci USA 91:12976
77. Irback A, Peterson C, Potthast F (1996) Proc Natl Acad Sci USA 93:9533
78. Govorun EN, Ivanov VA, Khokhlov AR, Khalatur PG, Borovinsky AL, Grosberg AY (2001) Phys Rev E 64:040903(R)
79. Shakhnovich EI (1994) Phys Rev Lett 72:3907
80. Shakhnovich EI (1996) Folding & Design 1:R50
81. Shakhnovich EI (1997) Curr Opin Struct Biol 7:29
82. Pande VS, Grosberg AY, Tanaka T (1997) Biophys J 73:3192
83. Pande VS, Grosberg AY, Tanaka T, Rokhsar DS (1998) Curr Opin Struct Biol 8:68
84. Sfatos CD, Shakhnovich EI (1997) Physics Reports 288:77
85. Grosberg AY (1997) Uspekhi Fizicheskih Nauk 167:129 (in Russian)
86. Kuhlman B, Dantas G, Ireton GC, Varani G, Stoddard BL, Baker D (2003) Science 302:1364
87. Wang G, Kuroda K, Enoki T, Grosberg A, Masamune S, Oya T, Takeoka Y, Tanaka T (2000) Proc Natl Acad Sci USA 97:9861
88. Tanaka T, Enoki T, Grosberg AY, Masamune S, Oya T, Takaoka Y, Tanaka K, Wang C, Wang G (1998) Berichte der Bunsen-Gesellschaft Phys.Chem 102:1529
89. Takeoka Y, Berker AN, Du R, Enoki T, Grosberg AY, Kardar M, Oya T, Tanaka K, Wang G, Yu X, Tanaka T (1999) Phys Rev Lett 82:4863
90. Enoki T, Oya T, Tanaka K, Watanabe T, Sakiyama T, Ito K, Takeoka Y, Wang G, Annaka M, Hara K, Du R, Chuang J, Wasserman K, Grosberg AY, Masamune S, Tanaka K (2000) Phys Rev Lett 85:5000
91. Ito K, Chuang J, Alvarez-Lorenzo J, Watanabe T, Ando N, Grosberg AY (2003) Prog Polym Sci 28:1489
92. Khokhlov AR, Khalatur PG, Ivanov VA, Chertovich AV, Lazutin AA (2002) In: Mac Kernan D (ed) Challenges in molecular simulations (SIMU Newsletter), Lyon: CECAM 4:79
93. Amit DJ, Gutfreund H, Sompolinsky H (1985) Phys Rev A 32:1007
94. Fink T, Ball R (2001) Phys Rev Lett 87:198103
95. Pande VS, Grosberg AY, Tanaka T (1994) J Chem Phys 101:8246
96. Pande VS, Grosberg AY, Tanaka T (1995) Macromolecules 28:2218
97. Pande VS, Grosberg AY, Tanaka T (1995) J Chem Phys 103:9482

98. Du R, Grosberg AY, Tanaka T (1998) Folding & Design 3:203
99. Sauer RT (1996) Folding and Design 1:R27
100. Davidson A, Sauer R (1994) Proc Natl Acad Sci USA 91:2146
101. Borovinskiy AL, Grosberg AY (2003) J Chem Phys 118:5201
102. Khokhlov AR, Khalatur PG (1998) Physica A 249:253
103. Khalatur PG, Ivanov VA, Shusharina NP, Khokhlov AR (1998) Russ Chem Bull 47:855
104. Khokhlov AR, Khalatur PG (2004) Curr Opin Solid State Mater Sci 8:3; Khokhlov AR, Khalatur PG (2005) Curr Opin Colloid Interface Sci 10:22
105. Khokhlov AR, Berezkin AV, Khalatur PG (2004) J Polym Sci Part A: Polym Chem 42:5339
106. Khalatur PG, Berezkin AV, Khokhlov AR (2004) Recent Develop Chem Phys 5:339
107. Khokhlov AR, Grosberg AY, Khalatur PG, Ivanov VA, Govorun EN, Chertovich AV, Lazutin AA (2001) In: Broglia RA, Shakhnovich EI (eds) Proceedings of the International School of Physics, Enrico Fermi, Course CXLV: Protein folding, evolution and design, IOS Press, Amsterdam, p 313
108. Kriksin YA, Khalatur PG, Khokhlov AR (2002) Macromol Theory Simul 11:213
109. Kriksin Yu, Khalatur P, Khokhlov A (2003) Macromol Symp 201:29
110. van den Oever JMP, Leermakers FAM, Fleer GJ, Ivanov VA, Shusharina NP, Khokhlov AR, Khalatur PG (2002) Phys Rev E 65:041708
111. Khalatur PG, Khokhlov AR, Mologin DA, Reineker P (2003) J Chem Phys 119:1232
112. Zherenkova LV, Khalatur PG, Khokhlov AR (2003) J Chem Phys 119:6959
113. Mologin DA, Khalatur PG, Khokhlov AR, Reineker P (2004) New J Phys 6:133
114. Zheligovskaya EA, Khalatur PG, Khokhlov AR (1999) Phys Rev E 59:3071
115. Starovoitova NY, Khalatur PG, Khokhlov AR (2004) In: Skjeltorp AT, Belushkin AV (eds) Forces, Growth and Form in Soft Condensed Matter: At the Interface between Physics and Biology. NATO science series: II: Mathematics, Physics and Chemistry, vol 160, Kluwer, Dordrecht
116. Starovoitova NYu, Khalatur PG, Khokhlov AR (2003) Dokl Chem 392:242
117. Starovoitova NYu, Berezkin AV, Kriksin YuA, Gallyamova OV, Khalatur PG, Khokhlov AR (2005) Macromolecules 38:2419
118. Velichko YS, Khalatur PG, Khokhlov AR (2003) Macromolecules 36:5047
119. Zherenkova LV, Talitskikh SK, Khalatur PG, Khokhlov AR (2002) Dokl Phys Chem 382:23
120. Zherenkova LV, Khalatur PG, Khokhlov AR (2003) Dokl Phys Chem 393:293
121. Berezkin AV, Khalatur PG, Khokhlov AR, J Chem Phys 118:8049
122. Berezkin AV, Khalatur PG, Khokhlov AR, Reineker P (2004) New J Phys 6:44; Berezkin AV, Khalatur PG, Khokhlov AR (2005) Polymer Sci A 47:66
123. Berezkin AV, Solov'ev MA, Khalatur PG, Khokhlov AR (2004) J Chem Phys 121:6011; Berezkin AV, Solov'ev MA, Khalatur PG, Khokhlov AR (2005) Polymer Sci A 47:622
124. Kriksin YuA, Khalatur PG, Khokhlov AR (2005) J Chem Phys 122:114703
125. Virtanen J, Baron C, Tenhu H (2000) Macromolecules 33:336
126. Siu MH, Zhang G, Wu C (2002) Macromolecules 35:2723
127. Lozinsky VI, Simenel IA, Kulakova VK, Kurskaya EA, Babushkina TA, Klimova TP, Burova TV, Dubovik AS, Grinberg VYa, Galaev IY, Mattiasson B, Khokhlov AR (2003) Macromolecules 36:7308
128. Lifshitz IM (1973) J Exp Theor Phys 65:1100; See also Lifshitz IM (1994) In: Selected scientific works. Electronic theory of metals. Polymers and biopolymers. Nauka publishers, Moscow, p 226 (in Russian)
129. Peng CK, Buldyrev SV, Goldberger AL, Havlin S, Sciortino F, Simons M, Stanley HE (1992) Nature 356:168

130. Danilowicz C, Coljee VW, Bouzigues C, Lubensky DK, Nelson DR, Prentiss M (2003) Proc Natl Acad Sci USA 100:1694
131. Blumenfeld LA (1974) Problemy biologicheskoi fiziki (Problems of biological physics). Nauka, Moscow (in Russian)
132. Volkenstein MV (1977) Molecular biophysics. Academic Press, New York; Volkenstein MV (1983) General biophysics, vol 1 & 2. Academic Press, New York
133. Blumenfeld LA (2002) Reshaemye i nereshaemye problemy biologicheskoi fiziki (Solvable and insolvable problems of biological physics). URSS Publishers, Moscow (in Russian)
134. Rubin AB (1999) Biophysics, vol 1, Theoretical biophysics. The University publishing house, Moscow (in Russian); Rubin AB (2000) Biophysics, vol 2, Biophysics of cellular processes. The University publishing house, Moscow (in Russian); Rubin AB (1994) Lectures on Biophysics. The University publishing house, Moscow (in Russian)
135. Blumenfeld LA, Tikhonov AN (1994) Biophysical thermodynamics of intracellular processes: molecular machines of the living cell. Springer, Berlin Heidelberg New York
136. McClare CW (1971) J Theor Biol 30:1
137. Stange P, Mikhailov AS, Hess B (1998) J Phys Chem B 102:6273
138. Eigen M (1971) Naturwissenschaften 58:10
139. Grosberg AYu, Khokhlov AR (1989) In: Fizika v Mire Polimerov (Physics in the world of polymers). Nauka, Moscow (in Russian)
140. Finkelstein AV, Gutin AM, Badretdinov AYa (1993) FEBS Lett 325:23; Finkelstein AV, Badretdinov AYa, Gutin AM (1995) Proteins 23:142
141. Li H, Helling R, Tang C, Wingreen N (1996) Science 273:666
142. Dokholyan NV, Shakhnovich B, Shakhnovich EI (2002) Proc Natl Acad Sci USA 99:14132; Deeds EJ, Dokholyan NV, Shakhnovich EI (2003) Biophys J 85:2962
143. Xia Y, Levitt M (2004) Curr Opin Struct Biol 14:202
144. Gutin AM, Abkevich VI, Shakhnovich EI (1995) Proc Natl Acad Sci USA 92:1282
145. Abkevich VI, Gutin AM, Shakhnovich EI (1996) Proc Natl Acad Sci USA 93:839
146. Khalatur PG, Novikov VV, Khokhlov AR (2003) Phys Rev E 67:051901
147. Chertovich AV, Govorun EN, Ivanov VA, Khalatur PG, Khokhlov AR (2004) Eur Phys J E 13:15

Author Index Volumes 101–196

Author Index Volumes 1-100 see Volume 100

Subject Index

MIX
Papier aus verantwortungsvollen Quellen
Paper from responsible sources
FSC® C105338

If you have any concerns about our products,
you can contact us on
ProductSafety@springernature.com

In case Publisher is established outside the EU,
the EU authorized representative is:
Springer Nature Customer Service Center GmbH
Europaplatz 3, 69115 Heidelberg, Germany

Printed by Libri Plureos GmbH
in Hamburg, Germany